Talking Mathematics in School

Studies of Teaching and Learning

Talking Mathematics in School investigates the relationship between students' discussions about mathematics in K–12 classrooms and their mathematical understanding. Beginning with a linguistic and sociolinguistic review of what is known about connections between thought, language, and learning, Lampert and Blunk consider what this research suggests for the teaching and learning of mathematical ideas and discourse. A collection of studies from various disciplinary perspectives – set in elementary and secondary classrooms, a computer-supported tutorial, and a workplace interaction – examines the nature of mathematical talk and the roles of students, teachers, tasks, and environment in producing it. Three studies were conducted in the same classroom, applying finer and finer analytic lenses to the relationship between classroom culture and mathematical talk, with an emphasis on what the teacher does to initiate and maintain a culture that supports students' engagement in mathematical practice.

Magdalene Lampert received her doctorate in Education from Harvard University in 1981. After nearly 20 years of teaching at every level, from preschool through graduate school, Lampert joined the faculty of Education at the University of Michigan in 1994. She has published widely in the fields of research on teaching, teacher education, and mathematics education.

Merrie L. Blunk received her doctorate in Education and Psychology from the University of Michigan in 1996. Her research focuses on students' interpretations of scientific concepts and on the challenges for teachers in taking account of those interpretations in the course of instruction. Her most recent work has concentrated on developing schemes for examining what students learn during collaborative small group work in classrooms.

Learning in Doing: Social, Cognitive, and Computational Perspectives

General Editors
ROY PEA, *SRI International, Center for Technology in Learning*
JOHN SEELY BROWN, *Xerox Palo Alto Research Center*
JAN HAWKINS, *Center for Children and Technology, New York*

Computation and Human Experience
PHILIP E. AGRE

The Computer as Medium
PETER BOGH ANDERSEN, BERIT HOLMQVIST, and JENS F. JENSEN (eds.)

Understanding Practice: Perspectives on Activity and Context
SETH CHAIKLIN and JEAN LAVE (eds.)

Situated Cognition: On Human Knowledge and Computer Representations
WILLIAM J. CLANCEY

Cognition and Tool Use: The Blacksmith at Work
CHARLES M. KELLER and JANET DIXON KELLER

Situated Learning: Legitimate Peripheral Participation
JEAN LAVE and ETIENNE WENGER

Sociocultural Psychology: Theory and Practice of Doing and Knowing
LAURA M. W. MARTIN, KATHERINE NELSON, and ETHEL TOBACH (eds.)

The Construction Zone: Working for Cognitive Change in School
DENIS NEWMAN, PEG GRIFFIN, and MICHAEL COLE

Street Mathematics and School Mathematics
TEREZINHA NUNES, DAVID WILLIAM CARRAHER, and
ANALUCIA DIAS SCHLIEMANN

Distributed Cognitions: Psychological and Educational Considerations
GAVRIEL SALOMON (ed.)

Plans and Situated Actions:
The Problem of Human–Machine Communication
LUCY A. SUCHMAN

Mind and Social Practice: Selected Writings by Sylvia Scribner
ETHEL TOBACH, RACHEL JOFFEE FALMAGNE, MARY B. PARLEE,
LAURA M. W. MARTIN, and AGGIE KAPELMAN (eds.)

Sociocultural Studies of Mind
JAMES V. WERTSCH, PABLO DEL RIO, and AMELIA ALVAREZ (eds.)

Communities of Practice: Learning, Meaning, and Identity
ETIENNE WENGER

Learning in Likely Places: Varieties of Apprenticeship in Japan
JOHN SINGLETON (ed.)

Talking Mathematics in School

Studies of Teaching and Learning

Edited by
MAGDALENE LAMPERT MERRIE L. BLUNK

PUBLISHED BY THE PRESS SYNDICATE OF THE UNIVERSITY OF CAMBRIDGE
The Pitt Building, Trumpington Street, Cambridge CB2 1RP, United Kingdom

CAMBRIDGE UNIVERSITY PRESS
The Edinburgh Building, Cambridge CB2 2RU, UK http://www.cup.cam.ac.uk
40 West 20th Street, New York, NY 10011-4211, USA http://www.cup.org
10 Stamford Road, Oakleigh, Melbourne 3166, Australia

First published 1998

Printed in the United States of America

Typeset in Ehrhardt 11/13 pt. in AMS-TEX [FH]

A catalog record for this book is available from the British Library.

Library of Congress Cataloging in Publication Data

Talking mathematics in school : studies of teaching and learning /
edited by Magdalene Lampert, Merrie L. Blunk.
p. cm. – (Learning in doing)
Includes bibliographical references and index.
ISBN 0-521-62136-4 (hb)
1. Mathematics – Study and teaching. I. Lampert, Magdalene.
II. Blunk, Merrie L. (Merrie Lynn), 1965– . III. Series.
QA11.T27 1998
510′.71 – dc21 97-35236
CIP

ISBN 0 521 62136 4 (hardback)

Contents

Series Foreword

This series for Cambridge University Press is becoming widely known as an international forum for studies of situated learning and cognition.

Innovative contributions from anthropology; cognitive, developmental, and cultural psychology; computer science; education; and social theory are providing theory and research that seeks new ways of understanding the social, historical, and contextual nature of the learning, thinking, and practice emerging from human activity. The empirical settings of these research inquiries range from the classroom, to the workplace, to the high-technology office, to learning in the streets and in other communities of practice.

The situated nature of learning and remembering through activity is a central fact. It may appear obvious that human minds develop in social situations, and that they come to appropriate the tools that culture provides to support and extend their sphere of activity and communicative competencies. But cognitive theories of knowledge representation and learning alone have not provided sufficient insight into these relationships.

This series was born of the conviction that new and exciting interdisciplinary syntheses are under way, as scholars and practitioners from diverse fields seek to develop theory and empirical investigations adequate for characterizing the complex relations of social and mental life, and for understanding successful learning wherever it occurs. The series invites contributions that advance our understanding of these seminal issues.

Roy Pea
John Seely Brown
Jan Hawkins

Acknowledgments

This volume grew out of a conference on mathematics learning and communication at the Center for Research in Mathematical Sciences Education at the University of Wisconsin. This conference was designed to bring together scholars who study classroom discourse with researchers in mathematics education. During the conference, senior researchers at the Center, Elizabeth Fennema and Tom Carpenter, expressed an interest in collecting current research on the connection between "talking mathematics" and learning it in school. The papers in this volume are a result of that collection effort and the volume was prepared with support from the Center. We thank Elizabeth and Tom for the idea and the support. Among the participants in that conference, we would particularly like to acknowledge Mary Catherine O'Connor for initiating the boundary-crossing work that has made it possible for scholars from different "discourse communities" to examine issues of mutual interest. The chapter by O'Connor in this volume is a version of a working paper prepared for the conference attendees. The preparation of the book was supported in part by the National Center for Research in Mathematical Sciences Education through a grant from the Office of Educational Research and Improvement, United States Department of Education (grant no. R117G1002), and the Wisconsin Center for Education Research, University of Wisconsin – Madison.

Contributors

Merrie L. Blunk
University of Michigan
School of Education
Ann Arbor, Michigan

Paul Cobb
Peabody College of Education
Vanderbilt University
Nashville, Tennessee

Rogers Hall
School of Education
University of California
at Berkeley
Berkeley, California

Deborah Hicks
University of Delaware
College of Education
Newark, Delaware

Magdalene Lampert
University of Michigan
School of Education
Ann Arbor, Michigan

Kay McClain
Peabody College of Education
Vanderbilt University
Nashville, Tennessee

Rodney E. McNair
Peabody College of Education
Vanderbilt University
Nashville, Tennessee

Mary Catherine O'Connor
Boston University
School of Education
Boston, Massachusetts

Peggy S. Rittenhouse
College of Education
Michigan State University
East Lansing, Michigan

Reed Stevens
School of Education
University of California
at Berkeley
Berkeley, California

Peri Weingrad
University of Michigan
School of Education
Ann Arbor, Michigan

1 Introduction

Magdalene Lampert

Consider the collection of regularly shaped flat blocks shown in Figure 1.1. Can two of them be joined to make a hexagon? What is a hexagon anyway? And what does it mean to "join" two of these pieces?

A Sample of Mathematical Talk in School

To introduce the ideas in this volume, I begin with an extended example of mathematical talk from my fifth-grade classroom. This talk will not be analyzed here but will serve to raise the central ideas in the chapters in this book. What does mathematical talk have to do with learning? How do students learn to discuss mathematics? What do the words they use refer to? What does a teacher do in a mathematical discussion? What does classroom discussion have to do with the way mathematical problem solving proceeds outside the classroom?

My fifth-grade students – in groups of four or five – worked with sets of blocks like the ones pictured in Figure 1.1 (called tangrams) to find out if two of them could make a hexagon. And they talked about what they were doing. They disagreed about the "rules of the game": Can "join" mean "overlap"? Can you turn the blocks over? Does "two" mean two of the same shape or two different ones? Does a hexagon have to have equal sides? I did not answer these questions or resolve their disagreements. I encouraged them to ask their classmates for clarification and to talk about why different positions on these questions might make sense and what different assumptions would imply for the solution. As they moved the shapes around, they talked about relationships among the pieces and their attributes: There are two sets of triangles – two small, two larger – that are "the same." If you put two "same" triangles together, you can make a square, or a bigger triangle, depending on how you turn them. The side of the smaller triangle fits exactly on the side of the square, but what would you call the shape it makes? How many sides does it have?

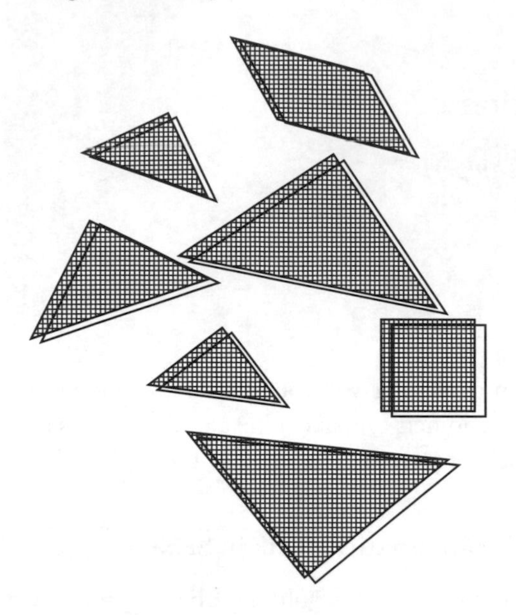

Figure 1.1. Tangrams

After about 20 minutes, I asked the students to stop their small group work and began a discussion intended to involve the whole class in considering whether one could make a hexagon with two tangram pieces. Awad tried to explain why it was impossible:

Lampert: OK. I just heard somebody say "it's impossible." [Some children whisper "yes" or "it is."] How could you prove that it's impossible to make a hexagon with two pieces? It's something that you have an intuition about; how could you prove it? Awad?

Awad: Um, first of all, a hexagon has six sides, you know, and then, like, if you take any of these shapes, you know, it won't make it; I mean it has to be like, see, all these lines are going this way and everything, but these don't do that.

Awad had his notebook propped open on his lap against the desk and glanced at it as he gestured with his pen and hand. I said something about what I thought Awad was getting at and asked if someone else who also thought it was impossible could expand on Awad's explanation. But, instead, Martin disagreed with Awad:

Lampert: OK. So the fact that a hexagon has six sides that you started out saying there, and the relationship between these shapes makes it hard to

Figure 1.2. Martin's hexagon

make it with two pieces. Could you, could somebody expand on what Awad said, or add, say your own thoughts.

Martin: Yeah, well, I know that you can. I just did it, it's not, it doesn't look like a hexagon, but it is, because it has six sides.

Everyone started talking at their tables. Several students addressed Martin directly. Shahroukh's voice could be heard above the din aggressively agreeing that Martin's figure did indeed have six sides and so it was a hexagon.

Shahroukh: Yeah! Yeah!
Lampert: OK.
Student: Let's see it.
Lampert: Now . . .
Students: Yeah, it has six sides . . .
Shahroukh: It has six sides, that's all you need!

Someone mentioned "six angles" and I asked Martin to hold up the figure so I and everyone else could see what he had done.

Students: Six angles . . . it has to have six angles . . . then . . .
Lampert: Let's see. Could you hold that up again, Martin, so that I could draw it on the board.

Martin held up the two pieces in one hand (see Figure 1.2). I drew his "solution" on the board. What I drew looked something like Figure 1.3. The class seemed disturbed by Martin's idea. There ensued a half-hour-long discussion, moderated by the teacher, in which assertions were made, using various kinds of evidence, about whether this was or was not a hexagon. The discussion veered into extended arguments about how to identify and measure the angles in the figure. There was a lot of disagreement about where the angles in the figure should be measured in relation to the "inside" and the "outside" of the figure, and this

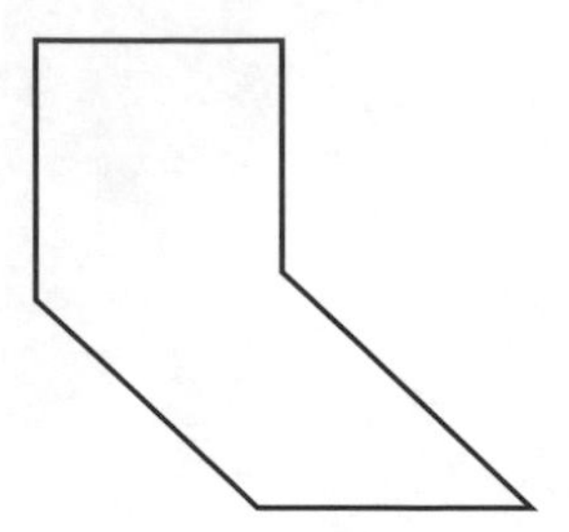

Figure 1.3. Drawing of Martin's hexagon

seemed to relate to how many angles the figure could be said to have, and thus to whether it was or was not a hexagon.

I asked Martin why he thought people might have been disturbed at his solution. He responded with a recognition that the figure that he had constructed in response to the puzzle was a little "strangely shaped" and it did not have equal sides. Does a hexagon have to have equal sides?

Lampert: Martin, why do you think people are a little disturbed by your idea?
Martin: Because, um, I, I usually think of a hexagon, I don't know if this is true of other people, but I usually think of a hexagon as having all *equal* sides, not being, you know, strangely shaped like that.

Much of the class seemed engaged with Martin's assertion, talking in their groups about whether he was right or wrong. I saw an opportunity to provoke some thinking and talking about important mathematical ideas, so I invited other students to comment on the assertion. Martin turned away from me to grab a dictionary and several students raised their hands. I asked a question about what Martin had said and several students indicated that they wanted to speak. I called on Dorota.

Lampert: OK and is that [i.e., equal sides] the only thing something has to have in order to be a hexagon? [pause, as Martin murmurs something inaudible] You're gonna look in the dictionary? Dorota?
Dorota: Six angles and six sides.

Next I asked a question that I thought would focus her and the rest of the class on an important geometrical idea and take us into more general mathematical territory. My intervention brought a chorus of student comments but not unanimity.

Lampert: OK, now Dorota said six angles and six sides. And what I would challenge you to say is, to figure out is, does every figure that has six sides also have six angles?
Students: No! No! Yes! . . . (many children talking at one time)

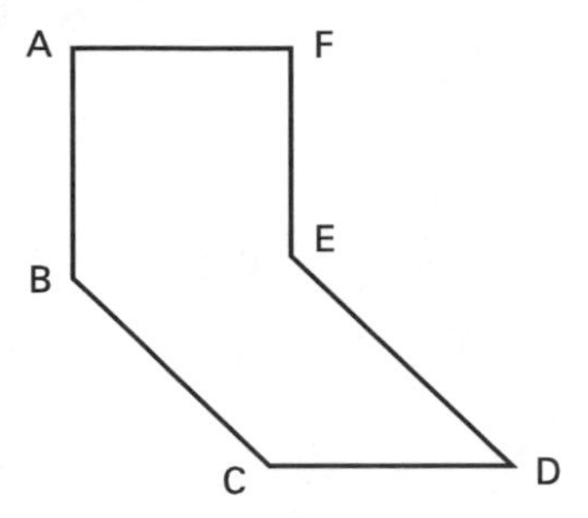

Figure 1.4. Labeled drawing

At this point I asked Martin to read the dictionary definition, but what he read did not address the disagreement on the table. He interpreted the dictionary definition as evidence that his assertion was correct, and he used it to make a distinction between regular and irregular hexagons.

Martin: I've found the definition . . . it says a plane figure having six sides, and six angles, and *that* [pointing to the board] is a hexagon. It's an irregular hexagon because it isn't shaped like, you know, the honeycomb figure that everyone pictures it as [Martin firmly closes the dictionary and returns it to the shelf].

Meanwhile Shahroukh asserted that the figure Martin had drawn was *not* a hexagon because it did not have six angles. As he spoke to me and the class, he hedged a bit and provoked me to focus him and the class on particular parts of the drawing on the board. I labeled some of the angles as a way of coordinating our communication while introducing a mathematical convention.

Lampert: Shahroukh, why at the beginning did you think it did not have six angles?

Shahroukh: Well, as soon as I saw it, I thought that two and two on the top and the bottom, that's four. And then, after I looked at it and then I saw two on the side, that's six . . .

Lampert: OK.

Shahroukh: . . . but, then, if you . . . well . . .

Martin: See, Dr. Lampert . . .

Lampert: I'm gonna label these also, so that we can talk about them a little better. [Teacher labels the figure as shown in Figure 1.4.]

Shahroukh did not miss a beat, incorporating the labels I had added in his attempt to explain his disagreement. Shahroukh continued:

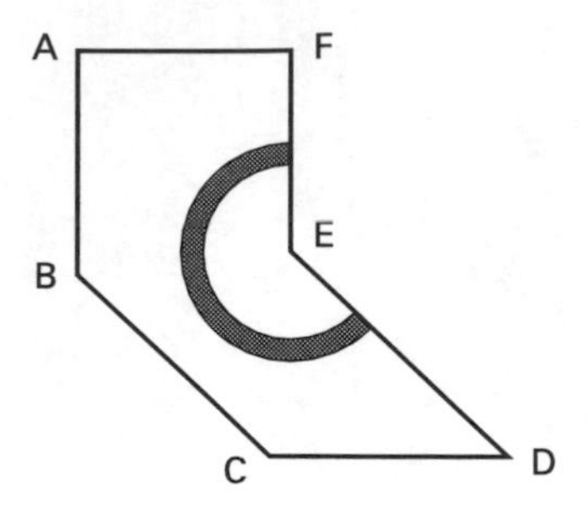

Figure 1.5. Angle E measured inside

Shahroukh: Why did I think it did not have six angles? Because when I saw it I thought of it as having it straight, you know, and then I thought there's C and D and A and F, and so that's four, and I started going, no, it doesn't have, it doesn't have six angles . . .

While he was addressing the class on what I considered to be an important matter, several other students were also talking to each other. I reminded the class that this was "large group." The routine for this part of the class, in place since the beginning of the year, was that only one person could be speaking at a time.[1]

Lampert: Wait, excuse me, we're having large group now and I don't think people are paying attention [to Shahroukh]. So you didn't think at first that B and E were angles?

From here, the discussion moved to a focus on angle E – was it inside or outside the figure? Arguments for which of these made sense were connected with questions about how to measure it. One student came up and gestured that angle E was as shown in Figure 1.5. Around the room, I could see by the ways other students were gesturing in their talk with one another that many disagreed. Another student came up and said you would need to measure it as in Figure 1.6, and a third said, no, it should be done "the smaller way," gesturing as in Figure 1.7. The discussion continued for another 20 minutes.

I present this small story of classwork here as an example of the kind of "math talk" that the authors of the studies reported in this book are seeking to understand. In the chapters that follow, the reader will find a variety of theoretical and analytical frames for addressing questions like: What are these students and this teacher talking about? (A juxtaposition of two blocks? A figure on the board? A set of mathematical abstractions?) What are the units of analysis we should use to answer this

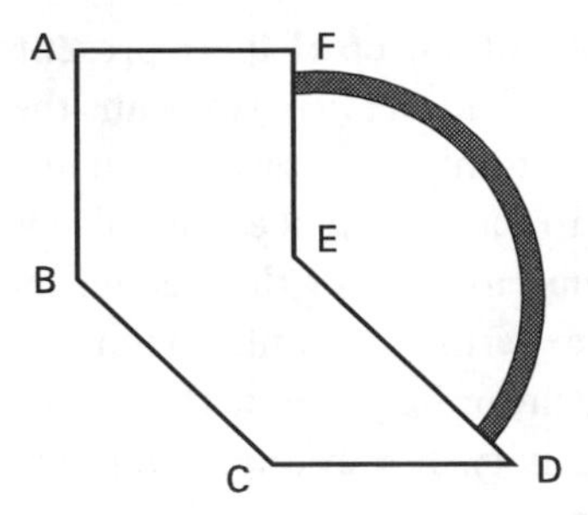

Figure 1.6. Angle E measured outside

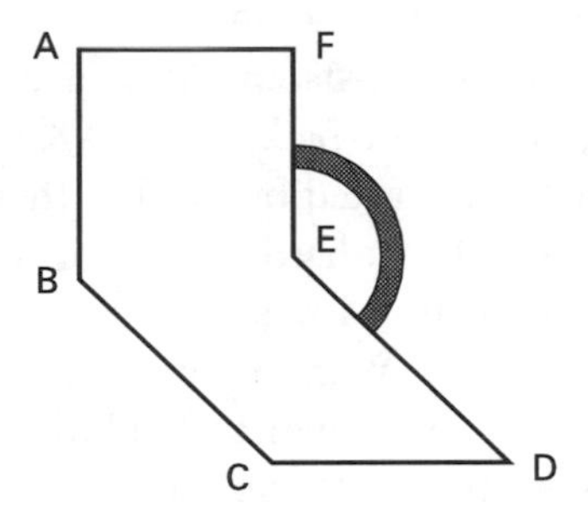

Figure 1.7. Angle E measured outside "the smaller way"

question? (The whole lesson? A thematic exchange? A single utterance?) What do speaker and spoken to do to check on whether they are talking about the same thing? Should they be talking about the same thing? What evidence exists that anyone is listening? Or understanding what is said? What are these students learning and how are they learning it? What is this teacher teaching and how is she teaching it?

Relating Talk and Learning

Much of what the fifth-grade students say in this geometry lesson might be called "inarticulate," and yet the teacher often seems to know what they are talking about, or acts as if they are talking about something mathematically coherent, as in the exchange between Awad and his teacher at the beginning of the story above. Some of my early insights into mathematical talk in classrooms actually came from recognizing the parallels between this teacher–student talk and the adult–child talk that goes on between caregivers and young children. There is some pointing and naming, some talking about objects that are present to both speakers, and some just talking. James Britton (1970) explains the importance of such primitive exchanges, not only as occasions when children learn the names of things and how to form sentences, but also as a process of organizing experience. As the caregiver or the teacher makes the child's utterances more explicit and experimentally fills in more of the situation than the child has expressed, she or he presents the ways in which society organizes time, space, and activity. For example, in the math lesson above, when Awad said, pointing to the figures in his notebook, "A hexagon has six sides, you know, and then, like, if you take any of these shapes, you know, it won't make it; I mean it has to be like, see, all these lines are going this way and everything, but these don't do that." I responded with a more mathematically coherent statement. I inserted the word *relationship* into my revoicing of what he had said: "O.K. So the fact that a hexagon has six sides that you started out saying there, and the relationship between these shapes makes it hard to make it with two pieces." With this expansion, the mathematics I was talking came closer to what would be heard outside the classroom among mature speakers, shaping what Awad and the others in the room might attend to.[2]

The contribution of Russian activity theory to the way we formulate the relationship between thought and language underscores this connection between words and worldview. This theory has been used by many scholars in the past 10 years to examine classroom discourse in all areas of the curriculum. Ideas like those I first encountered reading Britton in the 1970s are currently elaborated in reference to the work of Luria, Vygotsky, and Bakhtin, so that conversation and culture have become inseparable foci for investigation in classroom research (Newman, Griffin, & Cole, 1989; Moll, 1990; Hicks, 1996; Cazden, 1996). Activity theory has the child taking a central role in making language meaningful as it is acquired. The speaker of a new language is not a receiver of conventional definitions and ways of knowing, but an "appropriator." Like Martin with his strangely shaped hexagon, a student in school takes words from the culture of mathematics to make sense of something in experience, but also might push beyond conventional assumptions about how mathematical language can be used. The students responding to Martin reconstruct what the teacher has to offer as they negotiate what will stand as a "hexagon" and an "angle" in their talk. Deborah Hicks (1996, p. 108) boldly calls this process "learning": "As the child moves within the social world of the classroom, she appropriates (internalizes) but also reconstructs the discourses that constitute the social world of her classroom.

This creative process is what I would term learning." The practice of learning and the practice of teaching in school complement one another in the context of jointly constructed activity. Just as the child appropriates from the culture of the classroom, the teacher puts things out there to be appropriated, functioning as a partner in the conversation but with "a special mission and power" to ensure the classroom culture is rich in "offers, challenges, alternatives, and models, including languaging" (Bauersfeld, 1995, p. 283). In the example above, the activity is getting Martin's invention named and getting its angles identified, and the vehicles for this activity are talk, gesturing, and drawing. In these media, students and teacher are explaining, arguing, and demonstrating in ways that intend to convince the listener of a particular point of view.

Why Study Mathematical Talk in School?

Most classrooms are language rich, even those where the teacher does most of the talking. In a few classrooms, the teaching and learning of distinctly *mathematical* talk by having students engage in such talk is a deliberate pedagogical focus. The studies of teaching and learning included in this book have all been done in settings where the teacher's intention was that students should learn how to "talk mathematics." Why should we be interested in these classrooms? The work reported here occurs at the intersection of researchers' renewed attention to classroom discourse and the latest wave of reform in mathematics education. In the same year (1990) that I taught the lesson on tangrams, calls for the reform of mathematics in school classrooms began to be heard. In a broad-reaching proposal to change teaching in K–12 schools, the National Council of Teachers of Mathematics (1991, p. 3) asserted:

We need to shift

- toward classrooms as mathematical communities – away from classrooms as simply collections of individuals;
- toward logic and mathematical evidence as verification – away from the teacher as the sole authority for right answers;
- toward mathematical reasoning – away from merely memorizing procedures;
- toward conjecturing, inventing, and problem solving – away from an emphasis on mechanistic answer-finding;
- toward connecting mathematics, its ideas, and its applications – away from treating mathematics as a body of isolated concepts and procedures.

The message of this and other concurrent reform proposals is that students should learn from being more directly engaged in doing and talking about mathematics.

The emphasis and, indeed, even the substance of the new curriculum and pedagogy are formed from ideas about what mathematicians and people who use mathematics do when they are working on problems (NRC, 1988; Steen, 1990). They work in communities, they reason toward solutions and use logic to support their conclusions, they invent and make "educated guesses" called conjectures before they find solutions, and they connect elements of what they are doing with one another and with ideas outside the domain. If school lessons are to involve "communities" of learners doing this kind of mathematical work rather than individuals acquiring skills and remembering rules, classrooms will not be silent places where each learner is privately engaged with ideas. If students are to employ logic and mathematical evidence, they will need to compose speech acts or written artifacts that expose their reasoning. If they are to conjecture and connect, they will need to communicate. The idea that such classroom activities should be a goal of educational reform is based in part on current research in the learning sciences and research that examines the application of psychological theories to curriculum and instruction (see Romberg, 1990; Zarrinia, Lamon, & Romberg, 1987). The sorts of shifts in instructional practice advocated by the NCTM *Standards* are expected to produce increased understanding and improve practical competence.

How Shall We Study Mathematical Talk in School?

In order to measure whether certain kinds of classroom talk result in more and more desirable mathematical understanding, we need to investigate the kinds of curriculum and instruction that will support that talk. The studies in this book are an attempt to begin to understand these matters more clearly from the perspective of practice. The authors argue that we need a rich vision of what "talk" can be and a socioculturally complex concept of learning in order to relate the two. To develop this vision, they look carefully at a few classrooms in which teachers tried to work in the spirit of the reforms, where mathematical talk is primary among the activities that are structured to give students opportunities to learn. We acknowledge that these classrooms are not typical. The Third International Mathematics and Science Study, conducted in 1995, included a close look at teaching in classrooms in three of the participating countries – Japan, Germany, and the United States – and concluded that "the instructional habits and attitudes of U.S. mathematics are only beginning to change in the direction of implementation of mathematics reform recommendations" (National Center for Education Statistics, 1996, p. 47). In a sample of hundreds of classrooms at the eighth-grade level, lessons on the same topic were videotaped and transcribed. An analysis of this data found that in Japanese mathematics lessons, teachers typically pose a complex problem, students struggle with it, various students present ideas to the class, and the class discusses the various solution methods; in contrast, in the German and American classrooms, the lesson includes instruction by the teacher in a concept or skill, the teacher solving problems with the class, and students then practicing on their own (National Center for Education Statistics, 1996, p. 42).

We do have evidence that with appropriate support and structures in place, teachers can improve the quality of their mathematics instruction to build the capacity of students to think, reason, solve complex problems, and communicate mathematically (Brown, Stein, & Forman, 1996). The QUASAR project, working for 5 years with diverse populations in six urban school districts, has shown that this sort of improvement is both feasible and responsible. Silver, Smith, and Nelson (1995) provide vignettes of QUASAR classrooms in which students examine one another's reasoning and learn to express their mathematical ideas effectively; Stein, Grover, and Henningsen (1996) analyzed a representative sample of lessons in these classrooms and found that 75% of instructional episodes involved mathematical tasks intended to provoke students to engage in conceptual understanding, reasoning, or problem solving. Schools with an overwhelming majority of poor children, serving large subgroups whose first language is not English, have produced gains in students' proficiency with respect to mathematical understanding, problem solving, and communication (Silver & Stein, 1996).

Like the QUASAR classrooms, the teaching and learning environments studied by the researchers who have contributed to this volume are unusual, not only for their current scarcity but also because they have been created with uncommon access to many kinds of resources. They are what psychologist Ann Brown (1992) calls "design experiments."[3] Similar to prototypical work in fields like medicine and industry, they are created to be studied. The practices that are described have been constructed through a process of design and practical experiment informed by research on teaching, research on understanding, and research on the nature of mathematics. They focus on mathematical communication as well as on problem solving and quantitative reasoning. The research reported here is multidisciplinary, interdisciplinary, and often occurs on the boundaries between theory and practice. We examine talk from linguistic, sociolinguistic, mathematical, psychological, and anthropological perspectives. We look at mathematical talk in a nonclassroom learning setting by way of comparison.

Mathematical communication is neither a matter of curriculum nor a matter of instructional processes – it is both. In order to identify the elements of teaching and learning that occur at the intersection between subject matter transactions and classroom dynamics, it is necessary to give simultaneous attention to the intellectual content that is the focus of teacher–student interactions and to the social processes involved in constructing these interactions, and to ground analysis in the events that occur in classrooms (Erickson, 1982). At one level, the lesson described above is about mathematical vocabulary: what to call shapes and numbers. How is vocabulary taught and learned so as to make it usable? On another level, it is about learning how and when to express reasoning, disagreement, justification, exemplification. How are these discourse forms taught and learned? What is the relationship between learning vocabulary and learning discourse forms?

The studies in this book are divided according to whether they highlight doing and learning mathematical talk or teaching mathematical talk. By dividing up the territory in this way, we do not mean to suggest that these practices can be understood in isolation from one another. All of the chapters include descriptions of activities that integrate elements of doing, learning, and teaching. It is only in the choice of analytic lens that the work differs.

Notes

1. This mathematics class met for one hour, 4 to 5 days each week. Every day, we began with individual work for a few minutes. Then we had "small group" time during which students worked on problems with four or five others for about half an hour, followed by a half hour or so of teacher-moderated "large group" discussion. The social structure of these lessons was similar to that described in Wood, Cobb, and Yackel (1993).
2. The analogy between the acquisition of a disciplinary language like mathematics in the classroom and primary language acquisition in early childhood is a complex one. See Gee (1996) and O'Connor and Michaels (1993) for an application of this idea to classroom discourse.
3. Other such experiments with mathematical talk are reported by Silver and Stein (1996); Wood, Cobb, and Yackel (1993); and Forman (1995).

References

Bauersfeld, H. (1995). Language games in the mathematics classroom: Their function and their effects. In P. Cobb and H. Bauersfeld (Eds.), *The emergence of mathematical meaning: Interaction in classroom cultures*, pp. 271–291. Hillsdale, NJ: Erlbaum.

Britton, J. (1970). *Language and learning.* Hammondsworth: Penguin Books.

Brown, A. (1992). Design experiments: Theoretical and methodological challenges in creating complex interventions in classroom settings. *Journal of the Learning Sciences, 2* (2), 141–178.

Brown, C. A., Stein, M. K., & Forman, E. A. (1996). Assisting teachers and students to reform the mathematics classroom. *Educational Studies in Mathematics, 31* (1–2), 63–93.

Cazden, C. (1996). Selective traditions: Readings of Vygotsky in writing pedagogy. In D. Hicks (Ed.), *Discourse, language, and schooling,* pp. 165–185. New York: Cambridge University Press.

Erickson, F. (1982). Classroom discourse as improvisation: Relationships between academic task structure and social participation structure in lessons. In L. C. Wilkinson (Ed.), *Communicating in the classroom.* Madison, WI: University of Wisconsin Press.

Forman, E. (1995). Learning in the context of peer collaboration: A pluralistic perspective on goals and expertise. *Cognition and Instruction, 13,* 549–564.

Gee, J. (1996). Vygotsky and current debates in education: Some dilemmas as afterthoughts to *Discourse, language, and schooling.* In D. Hicks (Ed.), *Discourse, language, and schooling,* pp. 269–282. New York: Cambridge University Press.

Hicks, D. (1996). Contextual inquiries: A discourse-oriented study of classroom learning. In D. Hicks (Ed.), *Discourse, language, and schooling,* pp. 104–144. New York: Cambridge University Press.

Moll, L. (Ed.). (1990). *Vygotsky and education: Implications and applications of sociohistorical psychology.* New York: Cambridge University Press.

National Center for Education Statistics. (1996). *Pursuing excellence* (NCES 97–198). Washington, DC: U.S. Government Printing Office.

National Council of Teachers of Mathematics. (1991). *Professional standards for teaching mathematics.* Reston, VA: National Council of Teachers of Mathematics.

National Research Council. (1988). *Everybody counts: A report to the nation on the future of mathematics education.* Washington, DC: National Academy Press.

Newman, D., Griffin, P., & Cole, M. (1989). *The construction zone: Working for cognitive change in school.* New York: Cambridge University Press.

O'Connor, M. C., & Michaels, S. (1993). Aligning academic task and participation status through revoicing: Analysis of a classroom discourse strategy. *Anthropology and Education Quarterly, 24,* 318–335.

Romberg, T. (1990). Evidence which supports NCTM's Curriculum and Evaluation Standards for School Mathematics. *School Science and Mathematics, 90* (October, 1990), 466–479.

Silver, E. A., Smith, M. S., & Nelson, B. S. (1995). The QUASAR project: Equity concerns meet mathematics education reform in the middle school. In W. G. Secada, E. Fennema, & L. B. Adajian (Eds.), *New directions in equity in mathematics education,* pp. 9–56. New York: Cambridge University Press.

Silver, E. A., & Stein, M. K. (1996). The QUASAR project: The "revolution of the possible" in mathematics instructional reform in urban middle schools. *Urban Education, 30* (4), 476–521.

Steen, L. (1990). *On the shoulders of giants: New approaches to numeracy.* Washington, DC: National Academy Press.

Stein, M. K., Grover, B. W., & Henningsen, M. A. (1996). Building student capacity for mathematical thinking and reasoning: An analysis of mathematical tasks in reform classrooms. *American Educational Research Journal, 33* (2), 455–488.

Wood, T., Cobb, P., & Yackel, E. (1993). The nature of whole class discussions. In *Rethinking elementary school mathematics: Insights and issues,* Monograph. Reston, VA: National Council of Teachers of Mathematics.

Zarrinia, E. A., Lamon, S., & Romberg, T. (1987). *Epistemic teaching of school mathematics* (Program Report 87–3). Madison: School of Mathematics Monitoring Center, University of Wisconsin School of Education.

Part I

Doing and Learning Mathematical Talk

We begin this volume with an overview of the domain by Mary Catherine O'Connor investigating what might be meant by doing and learning mathematical "discourse" in school. O'Connor considers mathematics as a "practice" with its own norms for doing and talking, and she looks at how these norms might differ from those children learn in families and with peers. O'Connor considers a variety of "protoforms" of mathematical talk that are found in children's out-of-school speech behaviors. From this perspective, she raises fundamental questions about the transfer of capacities developed in the context of informal talk at home to mathematical work in school. O'Connor's analysis suggests several reasons why students who are able to argue and reason and use appropriate quantitative words at home might have difficulty importing these abilities into the social setting of the classroom. Her review provides a backdrop for research on what and how children need to *learn* if they are to "talk mathematics" in school. These questions are taken up in the rest of the studies in this book.

Chapters 3 and 4 are complementary analyses of the ways in which students do and do not connect mathematical talk with talk about actions on objects. Kay McClain and Paul Cobb draw on theories of mathematical learning and development to observe that the scenarios within which students work on school problems (i.e., the "stories" in story problems) make a significant difference in their capacity to make meaningful mathematical moves. Rodney McNair is also concerned with the connection between mathematical moves and moves in situations drawn from concrete problems. He analyzes three teaching and learning situations using problems to differentiate the roles that teachers and students play in creating mathematical frames for students' activities. He concludes that how students perceive the purpose of their work has an important influence on their capacity to "talk mathematics" in the context of their work on problems. In chapter 5, Reed Stevens and Rogers Hall take us out of the classroom to look at a tutoring session in algebra and an exchange between a junior and a senior civil engineer. As in the studies of classroom discussion, they are interested in what it is that makes talk *mathematical.* Their fine-grained analysis of discourse focuses on what in the context makes mathematical talk successful and when and why communication breaks down. They examine the role of objects and diagrams in talk about problems with an eye toward understanding how perception becomes "disciplined" to new ways of seeing. Their analysis raises questions about the conventional distinction between "concrete" and "abstract" and provokes us to think in new ways about the transition from school to work.

2 Language Socialization in the Mathematics Classroom: Discourse Practices and Mathematical Thinking

Mary Catherine O'Connor

Implicit in the current consensus documents about mathematics education reform is an idea concerning the connection between communication about mathematics and actual mathematics learning. That idea – that somehow communication facilitates mathematics learning – remains nebulous: Rarely is there discussion of why "communication" should enhance "learning" or even of what we might understand communication to be. Some researchers seem to be concerned with communication in the sense of clear explanation, while others seem to be concerned with complex social practices embodied in discourse, practices that indicate membership in a discipline. Before providing a sketch of the contents of this chapter, I will consider two perspectives on the topic of mathematics and communication, as a way of locating the proposals to be developed below.

The Process–Product Tradition

Within the process–product tradition, "communication" can be understood most simply as "talk about (mathematics)," and "learning" can be interpreted in terms of individuals' performance on assessments of their understanding. The idea introduced above then emerges in an obvious and seemingly testable form: More talk about mathematics enables more learning about mathematics. Most of the studies in the process–product tradition suggest that such a relationship exists. They find reliable effects for certain types of instruction, coded in terms of verbal interaction or discourse. For example, Webb (1991) carried out a study in which she examined the results of seventeen previous studies that explored peer interaction and achievement in small groups. The studies she cites include a wide variety of mathematical topics and grade levels (from problems on time and money in the 2nd grade to algebra in the 11th grade). She finds two strong relationships between language-mediated peer interaction and achievement: (a) low achievement correlates with receiving nonresponsive feedback (e.g., only being told the correct answer by one's peers, with no further information) and (b) high achievement correlates with the behavior of giving "elaborate explanations" to one's teammates.

Webb interprets this result as indicating that instructional interventions should be designed to teach students how to give each other elaborate explanations. However, she cites a study that did just that and found no effects on achievement (Swing & Peterson, 1982). The results of just one study that involved only a two-session training program with follow-up support of course do not invalidate Webb's proposal, but they do underline just how little is known about why such interactions might correlate with achievement. Webb does not address the possibility that students who are able to give such elaborate explanations show up as high achievers due to other characteristics, characteristics that might explain both the elaborate explanations and the high achievement levels. She seems to have a tacit hypothesis that the delivery of elaborate explanations itself enhances the explanation giver's achievement in a directly causal fashion. If we had a better idea of what those interactions consisted of, at a microanalytic level, we might be better able to understand why they correlate with higher achievement – What kinds of abilities and knowledge do they reflect? Will classroom discourse of some kind make such abilities and knowledge available to more students?

To answer this question, we must have more precise ways of characterizing the communication about mathematics that we wish to understand. *Communication* can be understood in terms of dyadic interaction, or in terms of higher level participant structures. There are studies in the process–product tradition claiming the beneficial effects of large group discussion on a variety of intellectual and cognitive outcomes. However, as Gray (1993) points out, these are largely uninterpretable due to the lack of a clear definition of what constitutes *group discussion.* Gray attempts to address this by constructing a set of criteria for the identification of classroom large group discussion.

In some sense Hiebert and Wearnes (1993) are in the process–product tradition, although they are far more specific than most in showing exactly what sorts of practices they are interested in. Their description of four types of questions asked by teachers in second-grade classrooms gives examples of actual utterances and follows up in a general way on the consequences for students of these teachers' question-asking practices. In addition, their studies have provided strong evidence of a traditional sort that there is a relationship between classroom discourse and learning outcomes. They themselves acknowledge, however, how difficult it is to isolate the nature of the relationship between instruction and outcomes.

Sociocultural Theory of Mind

When we pose these same questions within a different tradition, the concepts of communication and learning appear in a different light. In the past twenty years, the notion has become widely influential that individual thought can be usefully understood as an internal conversation with interlocutors, the general dimensions of which originate in recurrent prior social interaction. The idea that "intramental" cognition derives in part from sociohistorically specific social practices and "intermental" interactions stems from the work of Vygotsky, Leontyev, and other Soviet theoriests, as well as G. H. Mead. Wertsch (1985, 1991; Wertsch & Rupert, 1993) argues that collective, intermental cognition can only be understood in terms of the social and cultural context in which a specific intermental activity is embedded. This context will necessarily then shape any intramental learning that results from the collective enterprise. In this framework, then, the study of individual learning leads necessarily toward a study of the social.

This groundwork has led some mathematics and science educators toward the study of social interaction as the ground out of which robust mathematical and scientific thinking may grow (Resnick, Levine, & Teasley, 1991; Cobb, Wood, & Yackel, 1993; Hatano, 1993; Wertsch, 1991 inter alia). Against this background, the following questions can be posed: How might classroom interaction – crucially language-mediated – provide an environment for the development of mathematical and scientific thinking and "habits of mind"? And how might discourse activities in classrooms, orchestrated by teachers and other experts, provide for the socialization and enculturation of the student, leading to the development of the self as mathematically capable?

Overview

In this chapter I will assume some version of the second tradition, considering how and why verbal communication in mathematics classrooms may deepen students' learning. In the view I will elaborate, mathematical intellectual practices are themselves complex social practices, not simply cognitive routines or behaviors. If we view mathematical thinking as a set of social practices, then it makes sense to ask how classroom discourse plays a part in the practices themselves and in the complex and lengthy process of socialization into those practices. Not surprisingly, this perspective will also lead us to consider how home-based discourse conventions do or do not transfer into the classroom mathematics setting.

Although I will be concerned with the transition from home to school, from the outset it is important to make clear that there is another transition that is fully as problematic. A number of writers have addressed the complex issue of the interface between the intellectual practices that constitute "school science" and "school math" and the practices found in the real worlds of working scientists and mathematicians (e.g., cf. the papers in Greeno & Goldman, in press). This problem will not be dealt with in this chapter at all, although it will rear its interesting head from time to time.

In the sections that follow, I first will examine a proposal for a set of "mathematical modes of cognition" – an attempt by members of a national consensus panel to spell out the kinds of targets that reform-minded mathematics educators must set their sights on. I will recast these as social practices in order to highlight the relationship between what we normally think of as cognitive capacities and the social and discursive grounds in which they must be maintained. I will then make a linkage between some of these mathematical sociocognitive practices and the discourse practices that play major and minor parts in their existence. These discourse forms will be found to be special versions of forms that children engage in within their homes and communities, and I will discuss some of these "protoforms" of mathematical and scientific discourse practices.

In the context of looking at particular examples, the question will arise whether we can or should expect home-based discourse practices to afford students a "platform" from which to begin to participate in those practices characteristic of school mathematics. In the process of discussing several examples, I will attempt to clarify the relationships (or lack thereof) between home-based and school-based discourse practices, and how we might begin to explore this more deeply. I will at times question whether competence in discourse practices "transfers" to new domains. This should be understood as a transfer in the everyday sense, meant to evoke the incorporation of old skills into new settings. Lave's and Walkerdine's arguments concerning the problematic nature of the construct "transfer" within psychology are persuasive in my view and are not challenged here.

Mathematical Cognition and Discourse Practices

To connect our consideration of discourse practices and mathematical behavior, we must start with some quite general statements about what these terms mean. First I will discuss some general "habits of mind" and intellectual practices that have been proposed by various observers to be central components of mathematical behavior, the kind of mathematical behavior we want our students to participate in spontaneously, voluntarily, and powerfully. I will then suggest that the development of these proposed cognitive entities entails facility in certain social practices – practices that are at least partially constituted in discourse-based interactions.[1]

Mathematical Modes of Cognition

How might one go about characterizing the general cognitive practices of mathematics, the "diverse discipline that deals with data, measurements, and observations from science; with inference, deduction, and proof; and with mathematical models of natural phenomena, of human behavior, and of social systems" (NRC, 1989, p. 31)? The nature of mathematical knowledge, its creation and transmission, is far too vast and controversial for discussion in this paper (but see Restivo, Van Bendegem, & Fischer, 1993; Putnam, Lampert, & Peterson, 1990; and Lampert, 1990, for some directions). For the purposes of this paper I will adopt a pedagogically oriented, pragmatic perspective, one that is suggested by a consensus document of the National Research Council, drawing on the views and research of mathematicians, educators, education psychologists, statisticians, scientists and engineers, teachers from elementary through graduate schools, state and local administrators and members of government, and leaders of groups representing parents, business, and industry (see NRC, 1989, and references therein). The work of the NRC, in concert with the efforts of the National Council of Teachers of Mathematics, led to a set of curricular and pedagogical standards that were operationalized with unprecedented scope and detail (NCTM, 1989).

The NRC report lists the following six activities as central exemplars of "mathematical modes of cognition," in straightforward statements about the kinds of behaviors and activities that reform classroom instruction should foster.

> Modeling – representing worldly phenomena by mental constructs, often visual or symbolic, that capture important and useful features

> Optimization – finding the best solution (least expensive or most efficient) by asking "what if" and exploring all possibilities
>
> Symbolism – extending natural language to symbolic representation of abstract concepts in an economical form that makes possible both communication and computation
>
> Inference – reasoning from data, from premises, from graphs, from incomplete and inconsistent sources
>
> Logical analysis – seeking implications of premises and searching for first principles to explain observed phenomena
>
> Abstraction – singling out for special study certain properties common to many different phenomena
>
> (NRC, 1989, p. 31)

Another consensus document (AAAS, 1993) discusses the "habits of mind" that underlie robust scientific and mathematical problem-solving behaviors in the widest sense of that term: the ability and, crucially, the *inclination* to solve problems involving science, mathematics, and technology both in school and throughout adult life.

When we talk about these phenomena as "modes of cognition" and "habits of mind," they appear distant and monolithic: It is not easy to see how a student might come to master them. If, however, we take the Vygotskyan view that these modes of cognition emerge on the intramental plane from recurrent interactions that take place between actors on the intermental plane, then we can begin to envision what it might mean for classroom communication to be structured to support their development. This envisioning requires that we understand these modes of cognition as embedded in and made up of complex sociocognitive practices. Although experts engage in these practices in solitary as well as collaborative situations, the acquisition of these abilities takes place first on the intermental plane, in the sociocultural environments of social interaction.[2]

What Is a Social Practice?

If we entertain the Vygotskyan idea that forms of discourse become forms of thinking, what kind of research in the mathematics classroom is implied? Centrally, the specific nature of the linkages between social practices and cognitive or intellectual practices or "habits of mind" must be spelled out. Such accounts will necessarily be highly particular, indexed to specific classrooms and to specific aspects of mathematical thinking. Generality in such research is often an obstacle to understanding. Scribner and Cole's (1981) famous study of the consequences of literacy among the Vai provides a partial model for approaching this question.

Scribner and Cole developed the idea of literacy as more than the technology and skills of reading and writing. In opposition to literacy theorists who claimed that simply acquiring the cognitive skills of reading and writing automatically results in pervasive intellectual consequences, Scribner and Cole's "practice theory of literacy" views the cognitive skills of literacy as "intimately bound up with the nature of the practices that require them" (Cole & Scribner, 1975, p. 237). For example, they studied Liberians who were literate in the Vai language, but were otherwise unschooled; these people used the Vai writing system primarily to write personal letters. The standard theory of literacy at the time would have predicted that these Vai literates would have a wide variety of enhanced intellectual capacities. However, they performed no differently than nonliterates on the range of "logical and analytical thinking skills" often claimed to result from basic literacy (but really attributable, Scribner and Cole argue, to formal schooling). On the other hand, their immersion in letter writing did have cognitive consequences: It provided them with an enhanced ability to describe objects or events to those who had not witnessed them firsthand, an ability that Scribner and Cole were able to demonstrate in a testing situation. They argue that "in order to identify the consequences of literacy, we need to consider the specific characteristics of specific practices. And, in order to conduct such an analysis, we need to understand the larger social system that generates certain kinds of practices (and not others) and poses particular tasks for these practices (and not others)" (Cole & Scribner, 1975, p. 237).

What is meant here by "a practice"? Scribner and Cole are more explicit about this term than most:

> By a practice we mean a recurrent, goal-directed sequence of activities using a particular technology and particular systems of knowledge. We use the term "skills" to refer to the coordinated sets of actions involved in applying this knowledge in particular settings. A practice, then, consists of three components: technology, knowledge and skills. We can apply this concept to spheres of activity that are predominantly conceptual (for example, the practice of law) as well as to those that are predominantly sensory-motor (for example, the practice of weaving). . . . We may construe them more or less broadly to refer to entire domains of activity around a common object (for example, law) or to more specific endeavors within such domains (cross-examination or legal research). Whether defined in broad or narrow terms, practice always refers to socially developed and patterned ways of using technology and knowledge to accomplish tasks. Conversely, tasks that individuals engage in constitute a social practice when they are directed to socially recognized goals and make use of a shared technology and knowledge system. (1981, p. 236)

When an individual has been effectively "inserted" into a recognized social practice, the individual will grow to participate in the embedded forms of reasoning and behaving, as these are called for by the social practice in question. In the case of the Vai, Scribner and Cole were able to show that the cognitive consequences of literacy are not absolute and monolithic, but instead are highly specific and derivable from the demands of particular literacy practices that incorporate reading or writing for specific and delimited purposes. Individuals' literacy "skills" – ways of reasoning and behaving – are driven by the particular literacy practices they habitually engage in.

Discourse Practices and Social Practices

How are classroom discourse practices instances of social practices? They are socially developed and patterned ways of using the tools of language and communication, along with specific knowledge of both communicative patterns and the subject matter under discussion, whether mathematics, science, or any other area of disciplinary study. While there are commonalities in the ways that many teachers use discourse in their classrooms, to some extent each classroom, as a social unit, has its own unique inventory of recurrent, patterned ways of pursuing its own unique goals and tasks. Viewed from the perspective of language socialization, these discourse practices are a means whereby experts (teachers) gradually induct novices (students) into the desired forms of thinking, reasoning, and valuing.[3]

With respect to this chapter, I will suggest that the mathematical and scientific modes of cognition as identified by the NRC are complex activities learned within specific social practices,[4] including, it is hoped, at least some discourse practices found within a classroom.[5] To clarify what I intend by this, consider one of the "modes of cognition" listed above: optimization – "finding the best solution (least expensive or most efficient) by asking 'what if' and exploring all possibilities." Although we may be used to thinking of activities like "optimization" as a purely mental, individual pursuit or mode of solitary cognition, in fact it may be that the development and use of this mode of cognition draws upon extensive participation in discourse practices: chains of verbal challenges and justifications, the adducing of evidence to answer a challenge, the verbal expression of speculation, and the explicit drawing out of consequences of potential courses of action.

Consider a successful solution to the following simple optimization problem:

In a photo emulsion factory, chemical process A provides the input to chemical process B, and both are required before the final testing of the emulsion fluid. Process A and process B cannot be carried out on the same machinery, but the testing process must take place on the same machinery used for process A. Assume that each process takes four hours and that one cycle of the three processes yields one batch of emulsion fluid. What is the maximal number of batches that can be derived in 16 hours?

One path to a solution might involve representing the machines and the processes in a drawing that indicates the course of time. The problem solver might draw the machine required to conduct process A and the machine required to conduct process B, including in the representation their existence through 16 hours of time, as one cannot easily answer the question unless one further represents the 4-hour blocks of time within this 16 hours (see Figure 2.1). Next, one can symbolically start process A on the first machine and let it run for 4 hours. The constraints that order process A and process B tell the solver that the first instance of process B cannot start until 4 hours into the 16 hours. After that, the testing must take place over the third 4-hour block, back on machine 1. This mathematically sound, yet nonoptimizing approach leads one to the conclusion that only one batch of emulsion fluid can be generated in the 16-hour period.

If the problem solver is approaching the problem within the optimizing "mode of cognition," however, what is different? The first move is a query: How many machines do I have to work with? The problem does not say, but one draws on the pragmatics of word problems to assume that

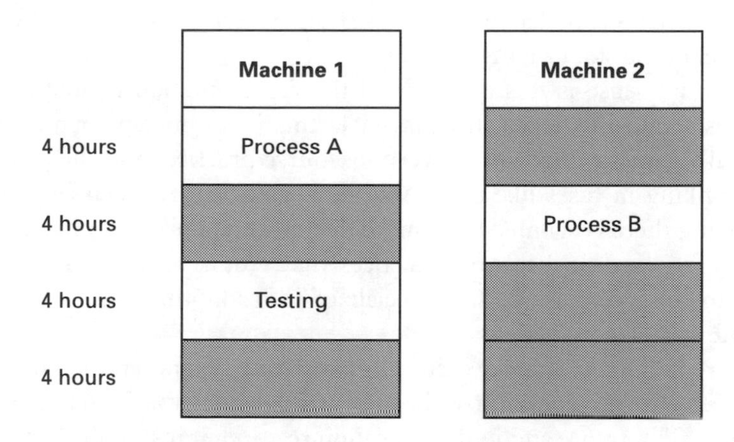

Figure 2.1. Solution no. 1

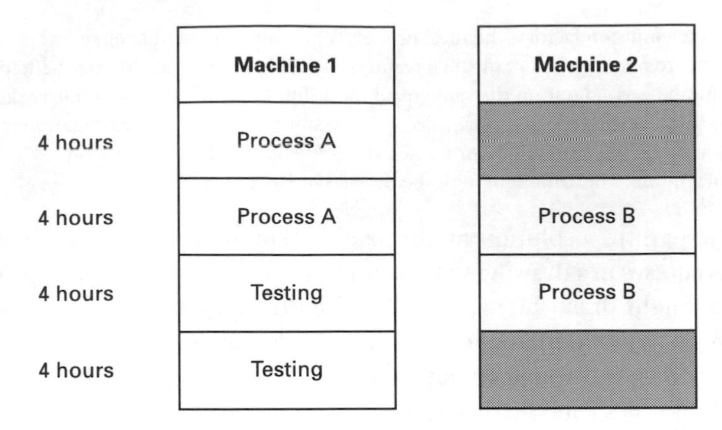

Figure 2.2. Solution no. 2

if more machines were available, the problem would have said so. An optimizer might start the process on machine 1, and would probably start to ask from the beginning "Can I do anything with the second machine during the first four hours?" The answer is no, that first batch of A output is needed to start the first batch of B. Nevertheless, the optimizer's query would be repeated about machine 1 during the second period of 4 hours. The answer now is yes, another batch of process A can start during that time. The query again recurs during the third period. When the testing is being conducted on machine 1, can anything happen on machine 2? This time the answer is again yes, a batch of process A is ready. Finally, that second batch of process B can be tested on machine 1 during the final time slot (see Figure 2.2).

In what sense is this querying of the representation related to challenges found in spoken discourse? Is this just a metaphor or is there actually some connection between discourse practices and the processes of thinking in cases like this? Within Vygotskyan and neo-Vygotskyan thinking there is claimed to be more than a metaphorical connection. By being involved in discourse practices that require one to challenge, to ask, to check, one is gradually socialized into adopting these as habits of mind.

A tremendous variety of culturally constituted events and activities that crucially feature speech have been described since the 1960s in anthropological literature and in the education research it inspired. These have been called *speech situations, speech events, speech activities,* and *discourse genres* (Hymes, 1962, 1974; Gumperz & Hymes, 1972). Representative examples are found in Gumperz and Hymes (1972), Hymes (1981), Sanches and Blount (1975), Bauman and Scherzer (1989), Scollon and Scollon (1981), Heath (1983), Michaels (1982), Goodwin (1990), and many others. Classroom discourse practices, like those of home and community, are composed of these sorts of entities: speech activities, events, and genres. It would be plausible to ask whether these discourse practices function as interactional scaffolds for cognitive activities. And if we view "cognitive modes" such as *optimization* as a complex activity, are there "protoforms" of this activity identifiable in the speech events and speech activities of the home?

Protoforms of Mathematical Discourse Practices

Before we consider classroom discourse forms that may support mathematical thinking, we might also consider what discourse practices outside of school may precede them. For example, the mathematical experiences of arguing, or of making a claim, or providing justification, or co-constructing a definition are abstract and ungraspable except as they take place within an activity or an event. That event must involve the learner. Experiential precursors (arguments outside of school, the provision of justification to parents and siblings, the struggles to name roles or objects in play) may provide the discourse "protoforms" that students could potentially build upon in the mathematical domain. When the learner enters new, more complex forms of school-based discourse, it may be that previous experiences are shaping expectations and guiding interpretations of what goes on.

Ochs and her colleagues (Ochs, Taylor, Rudolph, & Smith, 1992) make a striking claim about the origin of certain cognitive abilities within language-mediated events in the family, a special case of the socialization of cognition through language. She proposes that stories told at dinnertime in families are essentially protoforms of later theoretical thinking. In her account, "stories meet the explanatory criteria of theories in that storytellers posit at least one problematic event . . . which frames or recasts other narrated events as fitting into an explanatory sequence (e.g., cause and effect). . . . The second characteristic of all theories and potentially of stories as well is that their explanatory accounts are treated not as fact but as challengeable . . . either Initial Teller or Other Co-narrator . . . may treat the narrative exposition as but one possible *version* of experience and may call it into question and/or suggest an alternate explanation" (Ochs et al., 1992, p. 45).

This challenge to a narrative account is itself parallel, Ochs claims, to those found in "real" theories: "As in the realm of scientific and other scholarly debate, explanations in stories can be challenged at a factual level, at the level of methodology, and/or at the level of ideology" (p. 54). Ochs et al. argue that children hear and eventually participate in such challenges and rebuttals to experiences they have participated in, and thus they come to have facility with the discourse practices that eventually will be used to challenge and rebut aspects of scientific (and other theoretical) practice in school. Importantly, such family narrative contexts serve as a place where challenges to facts, method, and ideology can become *familiar* practices.

I have found a few exemplars in the literature of classroom discourse activities that provide parallels to Ochs's account. One is the classroom large group discussion led by Victoria Bill and examined by Bill in collaboration with Lauren Resnick and Mary Leer (Leer, Bill, & Resnick, 1993). Bill's interactions with her first, second, and third graders are full of exchanges like the following. The third-grade class has been discussing a problem in which Bill actually gave each student 15 Smarties (a candy tablet) to take home with them. Each child was allowed to eat two thirds of his or her Smarties. Bill has been asking them to tell her how many of their Smarties they got to eat. She has moved on to discuss the fractions two thirds and three thirds, with most questions built around representations on the board of three groups of 5 Smarties. Here she discusses what one student will get when she gets three thirds of 15.

1 VB: She's gonna get three groups. OK, she's gonna get three out of the three. Oh, how many does she get?
2 Students: (softly) Fifteen.
3 VB: Go ahead and count the whole thing. She (indicates student, Marnie) wants to count 'em. Go on up there and point to them. You don't have to count each one if you don't want, you can count however you want. Oh (hears a student and orients to that student). He says, count by *what*?
4 David: Five.
5 VB: (addressing student, Marnie, counting Smarties depicted on the board) David says you can count by five. Can you?
6 Marnie: Yes.
7 VB: Go ahead.
8 Marnie: One . . .
9 Students: (general sounds of consternation)
10 VB: (to class) She needs you. Do it again Marnie.
11 Students: (in unison) Five, ten, fifteen.

Here Vicki Bill orients the students to each other several times in different ways. In (3) through (5) she picks up on David's comment from his seat, made to no one in particular, that Marnie could count the Smarties by fives, she need not count each Smarty one at a time successively. She re-utters his contribution and queries Marnie, asking *her* to evaluate the correctness and current relevance of David's comment. (For further discussion of this example and the next, see O'Connor & Michaels, 1996.)

It is likely that both Marnie and David could engage in making claims and evaluating them in their lives outside the classroom. Virtually all speakers participate in such sequences in their daily lives: "It's too late for us to go to the park." "No it isn't, it's still light out." In this transcript, Bill is guiding students to engage in a familiar practice with an unfamiliar object: mathematical claims and directives.

Frequently, in examples like the following, Bill directly draws one student into further explicating the thought of another.

1 VB: You were allowed to eat two thirds of them. How many, did you get to eat?
[deletion of insertion sequences]
2 Ken: Because, you got to eat ten because there were fifteen, there were fifteen all Smarties in that bowl. And then you two thirds, that make, two out of the three groups, two out of the three groups, five plus five is three groups.
3 VB: Okay,
[deletion of insertion sequences]
so he said, open your notebooks . . . you have fifteen Smarties, and I said that you could have, two thirds, of, the fifteen, and I wanna know, how many, you were allowed to eat. Who can explain what Kenny did? . . . he said, David?
4 D: He said, five plus five, because (long pause)
5 VB: How did you know to do five and five though? Douglas?

By questioning the students and requiring them to verbalize what others have said and what they must be intending, Bill is socializing these students into the practice of representing the statements and propositional attitudes of others in the solution of a mathematics problem.

These same students are almost certainly able to repeat what others have said and to speculate as to what they are thinking or planning. As home-based discourse protoforms are experienced in new settings, with new purposes and objects, do children rely on their competence in the protoforms to bootstrap them into new relations with school objects and purposes? It is likely that the answer to this question is quite complex.

Recall the problem of optimization we considered above. Are there home-based protoforms of problem solving that might contribute to classroom facility in such cases? Imagine a real-life problem such as the following: "Your basketball coach says you need eight hours of sleep a night. You believe her, and besides you were really slow at practice today – too tired! However, you need to study for your math test tomorrow for at least two hours, you have basketball practice at 8:00 a.m., and you told your friend you'd watch a three-hour movie with him tonight. You live half an hour from school. What's your plan from the time school gets out today (3:00 p.m.) until 8:00 a.m. tomorrow?" The solution to this problem involves a bit of quantitative reasoning. You must lay down a template of hour slots and fit them to the incompressible quantity of time your activities require. This is the mathematical core, and it determines a huge number of possible solutions. You could sleep from 3:00 p.m. until 11:00 p.m., study until 1 a.m., travel home from school until 1:30 a.m., see the movie from 5 a.m. until 8 a.m., etc. But an equally huge number of real-world constraints make that possibility (and many others) not usable. As we solve this problem, we immediately constrain the set of mathematically given possibilities by the set of real-world constraints (one cannot sleep in the school; friends generally do not watch movies at 5 a.m.; one must go to the school for basketball practice; one may not fall asleep the moment one wishes, etc.).

In some ways, this problem is more difficult than the previous one about emulsion fluid batches. Theoretically, there are more degrees of freedom. Yet in this case, so many real-world constraints are brought to bear that it hardly feels like a mathematical problem at all. The feeling is exacerbated by the fact that the real-world constraints are so quotidian and overlearned that they don't seem to deserve mention. Only a few possible sequences come to mind at all. A child who generated even a large subset of all the possible ones would likely be judged to be quite odd.

Yet consider a developmentally earlier example of the same activity: A 6-year-old attempts to figure out how to fit several highly desired activities into a busy family Saturday, for example, visiting a favorite uncle to play computer games, going to the library, and watching a Mr. Bean rerun before going to late Saturday afternoon catechism class. To the caregivers' frustration and occasional amusement, the 6-year-old is liable to come up with a plan that completely ignores travel time, uncle's schedule, travel of the family car to sister's orthodontist appointment, and when lunch is going to fit in. How do these constraints (eventually) get learned? How does that 6-year-old learn to master the template of hours? In at least some cases, it seems safe to say, the child learns how to do this from weekly immersion in a family discourse practice of "scheduling talk."

Just as we can observe Ochs's family dinner talk as a precursor to theorizing, it is appealing to see early parallels to discrete math in family scheduling talk. But does the consistent verbal expression of speculation, and the verbal habit of querying possible scenarios, facilitate the incremental development of the "optimizing" habit of mind? Many students who can ably manage their desired string of activities throughout a busy day would wilt in the face of the relatively simple emulsion-scheduling problem. Why might this be?

Further Questions About "Transfer." Research on language socialization within the home clearly establishes that children learn to participate in complex social practices by gradually being inducted into those practices from their earliest days in the family. One might assume that speech activities learned at home can be easily transferred to school-based versions of the "same" activities. A student who can argue on the playground should be able to import that ability into the mathematics classroom. Experienced teachers know, however, that children may have full competence in these activities, and in their constitutive subactivities within their everyday lives, and may still not be able or willing to spontaneously engage in those practices in the domains of math and science. What might explain this lack of transfer?

The following anecdote exemplifies this issue and makes clear how difficult it may be to find the answers. I recently spent some time tutoring a bright high school freshman who had fallen behind in algebra for a number of good reasons having to do with catastrophic events in her home life. We were discussing a section of her textbook entitled "Counterexamples." The sample problem read, "Larry makes a conjecture that x^2 is greater than x for all values of x. Find a counterexample." My friend asked me what she was supposed to do. We talked about conjectures as "educated guesses" or hypotheses that could be checked out. I made sure she understood that Larry was conjecturing that no matter what value you choose for x, x^2 would always be greater than x. I then asked her whether she could check that out and see if it were true, or whether she could find some counterexamples, examples that would show the conjecture didn't always hold up. She promptly plugged in a few integers (all positive) and stated that yes, it was true. I asked her again whether she could think of any counterexamples, values for x that would show that the conjecture was not always true. She thought for about five seconds and said she couldn't think of any.

I then said I would comc up with a conjecture and ask her to find counterexamples. I said "Everybody who is in the hospital got there by eating a bad diet." She immediately came up with two good counterexamples from her experience. I accepted them.

Is this student unable to argue against a claim using specific examples – to "find counterexamples" as the book so telegraphically asks her to do? Clearly not. In the realm of her everyday experience she does this all the time. Then why did she show no inclination to actively and aggressively *search* for counterexamples in the case of the algebra problem? She could see that the conjecture would hold for all positive integers. Why didn't she push further and explore negative integers and fractions? She had the desire to find the answer, as evidenced by her behavior throughout our tutoring sessions; there were no "attitude problems." And she had the competence to carry out the necessary computations, as demonstrated by her ability to use negative numbers when prompted (immediately seeing that conjecture still held) and to use fractions when I explicitly gave her the command "try one fourth."

There are at least three classes of answers to the question of why she could find counterexamples in everyday life but seemed not to want to search for them in school mathematics. One type of answer involves the student's specifically mathematical knowledge and beliefs. It might be that she simply avoids fractions in general because she does not like to work with them. Therefore she would not spontaneously think about them in this problem because she avoids them in general. Or perhaps she doesn't think of fractional numbers as possible values of x, having a shallow and stereotyped understanding of the actions required to explore an equation with variables – "use whole numbers." Maybe she is responding to some belief about the nature of textbook problems in mathematics: The answer will usually be obvious, and one should not have to go far to find it out. That is, either each and every number will render the statement "x^2 is greater than x" to be true, or each and every number will render it false. One should not, according to this expectation, have to look at any more than one or two values of x. (These beliefs and attitudes are well attested by Schoenfeld [1989], among others.)

All of these explanations suggest that the problem has nothing to do with discourse practices but rather stems from an impoverished number sense, computational ability, or familiarity with algebraic objects. In this scenario, facility with discourse practices is a background factor, necessary but not sufficient to ensure that students will actually engage in those practices incorporating mathematical objects and relations.

Another class of explanations assumes that the student has the mathematical capability and commands the home-based version of the discourse activity, but she is still not able to perform the mathematical discourse activity – in this case, finding and verbalizing a mathematical counterexample. How could a student be conversant with the forms or protoforms of the discourse activity and yet not be able to put that facility into play with mathematical objects? Home-based discourse practices may include socioculturally specific constraints on performance for certain persons or topics. Heath, Ochs, Schieffelin, Gumperz, Schiffrin, and others have shown that "ways with words" and specific speech activities differ radically from community to community, and from culture to culture. The ways they vary are numerous. Within an arbitrarily chosen speech community a particular speech activity, say "verbal teasing," may be absent, or it may have any number of unpredictable social meanings. Communities will differ in what they consider appropriate topics for verbal teasing and in who they consider to be appropriate perpetrators and targets of verbal teasing.

For the purposes of this chapter, what hangs on these cultural (or even family-level) differences in language use conventions? If we consider a discourse practice as constituting the underpinnings of a developing mathematical behavior, then access to that practice, or beliefs about it, or culturally conventional restrictions on its use will all have consequences for the student's progress in mathematical thinking and communication – consequences that determine which students feel entitled and inclined to engage in the discourse practices required in reform-minded mathematics classrooms. Lampert, Rittenhouse, and Crumbaugh (1996) have begun to explore the different social meanings of argument in mathematics classes, but much more remains to be explored.

Ochs's (1991) description of verbal speculation in Samoa provides an example. She points out that cross-cultural research has shown that societies vary widely in what they consider to be the socially acceptable limits of speculation. Western European societies generally consider the mental contents of another person's mind to be an acceptable topic for speculation. The legal system, pedagogical practices, and general conversational etiquette in such societies allow for the probing and discussion of others' intentions and thoughts. Socialization practices vividly demonstrate this:

> That unclear mental states are an acceptable object of verbal speculation is made linguistically evident to children early in their lives. Dozens of times a day caregivers in middle-class mainstream communities explicitly guess at unclearly formulated thoughts of young children. The caregivers make a wild or educated guess at what these children may have in mind and ask them to confirm or disconfirm their hypotheses. . . . Other societies . . . strongly disprefer verbal speculation on what someone else might be thinking or feeling. In traditional Samoan households, interlocutors do not typically pose test questions nor do they engage in mind-reading games or riddles. . . . When Samoan caregivers hear an unclearly expressed thought of a young child, they do not engage the child in hypothesis testing vis-à-vis that thought. Rather, as in other societies, Samoan caregivers prefer to elicit a more intelligible reformulation of the thought – asking "What did you say?" for example – or to terminate the topic. (Ochs, 1991, p. 299)

Interactional demands of the home-based and school-based versions of an activity thus may differ, and they may differ further depending upon the social characteristics of the person involved. Consider the finding of counterexamples as a social practice. Finding counterexamples is a contentious activity: One has to actively search for examples that will defeat the claims of another. In everyday discourse, we are occasionally called upon to do this, but the question quickly emerges: How far is one entitled to go to find a counterexample? How hard should one appear to be searching for examples that shoot down the claims of another person? What kind of person aggressively and relentlessly goes after counterexamples that will show the claims of another to be false or wrongheaded? In our society, such behavior is likely to be negatively judged when one is a 14-year-old middle-class girl. This, then, is another possible explanation of the reluctance of my 14-year-old friend to find counterexamples to the conjecture given in the book. She is cognitively able to do so but has not adopted that practice into her repertoire of automatic behaviors, due to its social implications.

Finally, there is a third class of explanations, coming out of the work on situated cognition. The early work of Scribner and Cole would predict that, for example, engagement in home-based discourse practices of speculation and of querying possible scenarios would not, on its own, ineluctably result in high performance on tests of optimization in mathematical situations. The effect is not simply one of repetition and practice of certain discourse strategies or moves. In the social practice framework Scribner and Cole develop, such facilitation occurs only within meaningful, situated problem solving with a particular knowledge base, and the child must be inducted into the practice of speculating for a purpose – a mathematical purpose. In that way, the "habits" of hypothetical thinking required to optimize in mathematics develop through the recurrent, situated integration of mathematical knowledge and discourse practice. Similarly, Lave (1988) provides a number of examples that show how situations structure and select for specific means of reasoning, giving us many reasons to assume that discourse practices will not transfer into a new content domain without some very focused instruction, and perhaps not even then.

How then can one get a student like my young friend to spontaneously search beyond the positive integers for counterexamples? More generally, how can one socialize students to spontaneously engage in the kinds of discourse practices they already command – arguing, defining, speculating, challenging, modeling – with a new set of contents, mathematical objects and relations? If one attempts to achieve this result simply by increasing students' number sense and computational ability, by drill and practice in doing computations with fractions and negative numbers, is there a likelihood that they will then incorporate those facts and routines into higher order thinking activities? The research cited above suggests that increased ability to traverse the mathematical universe would involve both more robust mathematical knowledge and scaffolded practice carrying out everyday discourse routines while operating within the mathematical realm.

Mathematical Discourse Forms and Protoforms: Two Examples

I have sketched a complicated picture so far: the interwoven development of abilities and meta-abilities in mathematical thinking and in discourse practices. Complex mathematical thinking activities are presumed to be composed at least partially (and at least in the initial learning process) of discourse practices. These discourse practices may be found in the home-based experience of the child. Anecdotal and research-based evidence indicate these home-based protoforms will not transfer easily into the mathematical activities of school, although not much is known about this. So now I will return to the NRC list, seeking to describe two types of speech activities or practices that can plausibly be related to some of the more highly organized mathematical modes of cognition included there. In each subsection I lay out the target competencies, the protoforms that may constitute the beginnings of development of these competencies, and where possible, I present at least partial examples from published research on mathematics classrooms. I also discuss possible obstacles to transfer of protoforms.

Argument: Claims, Warrants, and Opposition

Frequently one encounters the observation that to think mathematically, students must be able to make public assertions about patterns they see – they must be able to make claims about relationships between mathematical entities (NCTM, 1989; NRC, 1989). They must be able to present various kinds of support for their positions, including explicit accounts of their reasoning and warrants for its legitimacy and correctness. Moreover, they must be able to generate these observations and assertions on their own, take ownership of the thinking that must be done, and break away from the belief fostered by much of the schooling process, that authority resides only in books and teachers. At the very least, they must develop the belief that even if authority frequently does reside legitimately in books and teachers, they as individuals are responsible for understanding and querying the sources of that authority. *Argumentation* encompasses all of these abilities and actions. In intermental contexts it is required for communication about all the "mathematical modes of cognition" cited above, and it is also clearly part of the intramental processes involved in *inference* and *logical analysis* as well.

Toulmin (1958), who is frequently cited in mathematics and science education research on discipline-based argumentation, suggests links between arguments in mathematics and science and those of everyday discourse. In his account of logical argument within the disciplines, he promotes the metaphor of jurisprudence:

> The rules of logic may not be tips or generalizations: they none the less apply to men and their arguments – not in the way that laws of psychology or maxims of method apply, but rather as *standards of achievement* which a man, in arguing, can come up to or fall short of, and by which his arguments can be judged. A sound argument, a well-grounded or firmly-backed claim, is one which will stand up to criticism, one for which a case can be presented coming up to the standard required if it is to deserve a favourable verdict. (Toulmin, 1958, p. 8)

The valid arguments characteristic of each discipline are seen not as a completely general phenomenon representing an ideal logic, but as uniquely grounded in some ways: "There is no explanation of the fact that one sort of argument works in physics, for instance, except a deeper argument also within physics" (Toulmin, 1958, pp. 258–259). In Toulmin's formulation there are two components to discipline-based argumentation: one is the knowledge and reasoning that is particular to a field, its peculiar logic, and the other is everyday communicative competence in the area of argument and dispute. This is to say that in learning to engage in mathematical argumentation, general "jurisprudential" moves will be combined with domain-specific kinds of claims, warrants, evidence, and inference.

So one eventual goal of mathematical enculturation is the ability to make and justify a claim, to present a well-grounded case, as it were. What are the precursors to this ability? What kinds of competencies must classroom discourse practices build upon? It is possible to trace this aspect of communicative competence to a point early in life, when the child begins to engage in oppositional argument.

Schiffrin (1985) makes a distinction between rhetorical arguments (those in which a speaker "presents an intact monologue supporting a disputable position") and oppositional arguments (those in which "one or more speakers support openly disputed positions"). As Goodwin (1990) and Eisenberg and Garvey (1981) report, oppositional arguments are fully controlled by very young children. The intricacies of their strategies in maintaining and building dispute sequences are beautifully traced out in Goodwin's ethnographic work. We might expect that rhetorical arguments would be a later kind of communicative competence, depending upon control of oppositional dispute strategies, as well as the ability to envision and anticipate the responses of a virtual interlocutor, or one who is not necessarily engaged in the dispute locally.

Orsolini and Pontecorvo (1992) explicitly draw the link between children's early communicative competence in dispute and the desired properties of "classroom discussion." They point out that "second-pair turns" (such as disagreeing or agreeing responses to previous turns) "tend to incorporate pragmatic and semantic features of the utterances to which they reply" (p. 116). Due to their connectedness to what precedes them, and due to their expectedness, second-pair turns may provide a particular micro-learning environment for the child. Orsolini and Pontecorvo hypothesize that "the sequential organization of talk brings about conversational expectations that may face the child with new communicative acts. In particular, when an adult aims to elicit children's explanations, talk can draw on the conversational expectations of dispute. In disputes, accounts and justifications are expected interactive moves" (p. 117). In this passage, Orsolini and Pontecorvo are claiming something similar to the proposal that opened this section: that children have competence in protoforms of certain discourse genres, and that these can form part of the base for their gradual expansion into more advanced varictics of communication.

Orsolini and Pontecorvo are not looking at mathematics classrooms, but their experimental study in a Roman preschool (children ages 5–6) provides a model for the current set of questions. Their aim was to produce discussion, in order to study the character of children's contributions. Their instructional aims were "facilitating children's topical talk and children's arguing." To produce discussion, they introduced several conditions (p. 118): (a) a collective experience (group reading, a lab activity, etc.) that was "structured to lead to more than a single solution. . . . Such an experience provides a shared universe of referents . . . and motivates subsequent talk"; (b) a period of teacher–pupil group discourse, designed to compare and evaluate different children's experiences; (c) disputes were allowed to arise without the teacher's attempting to control them; (d) turn-taking rules were changed, with the teacher not selecting the next speaker and limiting her own comments to rephrasing and repetitions; and (e) a change in the teacher's beliefs about how one learns in school.

I will focus here only on their findings about disputes. They found that even these very young children produced accounts and justifications during the disputes. In their view, these accounts and justifications were "dependent on the expectations raised by opposition" (p. 134). In other words, children know how to argue, and the maintenance of a dispute "forces the speaker to reconsider his or her own previous claim and to turn it into an articulate taking of a position." They conclude by suggesting that explanatory talk, of the type that educators want to promote, emerges out of the interaction of two different acts: justifying opposition and providing an answer to a teacher's request for explanation.

Orsolini and Pontecorvo focus on the ways in which children's preexisting interactional routines form a scaffold for more highly developed classroom argument. Hatano and Inagaki (1991) are also prominent developers of this tack. The Itakura method in science, studied by Hatano and his colleagues, tacitly exploits children's preexisting interactional routines by setting up a situation in which multiple possible outcomes create the conditions for discussion and for the taking of opposing positions. Hatano makes the point that as students take positions, they affiliate with each other in opposition to others. This partisanship is a lever that advances their thinking about the scientific phenomenon under consideration.

Pontecorvo and Girardet (1993) describe a group of 9-year-old students discussing and evaluating a Roman historian's judgments about the German populations in the fourth century. They argue that their example shows that children can practice and master the methodological and explanatory tools of a discipline – history, for example – in the context of autonomous collective discourse. They suggest that their analytic methods may make it possible eventually to "verify the evolution and/or the 'passage' of a reasoning structure from one child to another as an effect of the interaction with peers in appropriate learning environments. Indeed, the exchange concerns not only the appropriation of informational elements but also the acquisition of reasoning strategies, the core of which is given by the structures of justification" (p. 392).

Conflict Versus Oppositional Argument

Is it the case that where we find conflict, we will find increased learning? Webb (1991) reviewed several studies that explored a correlation between dissension or conflict in the group and achievement and reports that most found no effects or only marginal effects. She does, however, report one study whose results suggest that there is a complex relationship between verbal disagreements and achievement outcomes. Bearison, Magzamen, and Filardo (1986) found a nonlinear relationship between these factors. There are a few other pieces of work on small group interaction and mathematics learning that suggest that we should not equate open conflict with oppositional argument, although the two often go hand in hand. Joint problem solving may entail moments of oppositional argument that are not necessarily accompanied by "conflict." The study of oppositional argument should be carried out within the context of both conflictual and collaborative interaction. Cobb, Wood, Yackel, and their collaborators provide more detailed glimpses into group interaction than any of the studies cited by Webb. In Yackel, Cobb, and Wood (1991), the focus is on the process of small group work within cooperative learning, rather than on documenting, for example, size of effects associated with small group interaction time. Yackel et al. assert that the collaborative moments that can arise when students have the opportunity to work in small groups can be as intellectually useful as moments involving either cognitive or interpersonal conflict about a content issue.

As the children work together and strive to communicate, opportunities arise naturally for them to verbalize their thinking, explain or justify their solutions, and ask for clarifications. Further, attempts to resolve conflicts lead to both the opportunity to reconceptualize a problem and thus construct a framework for another solution method, and the opportunity to analyze an erroneous solution method and provide a clarifying explanation. . . . Opportunities for mathematics learning also arise when children attempt to reach consensus as they work together. In this situation, each is obliged to explain and justify his or her solution method to the partner and to listen to (as opposed to hear) the partner's explanation. (Yackel et al., 1991, pp. 401–402)

Interestingly, Yackel et al. also include instances in which there is no clear jointly focused work at all: At one level of description each interactant is simply running on a parallel but separate track. Yet the outcome shows mutual influence. Although we might expect that collaboration and opposition would be easy to spot and tell apart, this detailed work on small group interaction (see also Cobb, Wood, & Yackel, 1993) shows how complex and hard to recognize actual instantiations of collaboration and opposition may really be.

Unwanted Transfer: Social Image and Credibility

Vygotskyan and neo-Vygotskyan theories prompt us to pay attention to the cognitive potential of small group interactions like those described above. Forman and Cazden (1985) provide evidence that cognitive benefits do accrue to children who work together. Yackel et al. (1991) make a similar claim. In the research literature, both theory and empirical evidence seem to point in the direction of an instructional breakthrough. But within the classroom, the fragility of these arrangements is all too apparent. Lampert et al. (1996) show that many children are averse to argument in mathematics class. Students attach a negative meaning to argument in mathematics classes, by virtue of the fact that a negative social meaning would attach to arguing *outside* of class. It is not easy to extract a speech activity from all previous experience and change its essential meaning and purpose.

Here I will introduce another potential problem, where an undesirable transfer takes place. Oppositional argument outside a classroom takes into account the credibility, social standing, and prestige of the interlocutors. For the purpose of mathematics argument the identity of the student is supposed to be incidental: what is supposed to matter is the quality of the reasoning. Oppositional mathematical argument, some have argued, can develop best within classroom settings that foster real exploration of open-ended problems. However, just such settings often give rise to florid interpersonal conflict that is extremely difficult to tease apart from an ongoing, intensely waged exploration. A review of these ideas would be incomplete without at least one example of what is all too often the norm. The following example comes from field notes taken during a year-long observation of a sixth-grade mathematics class in an urban school. (All names have been changed; further discussion is given in O'Connor [1996].)

The small group observed was composed of two girls (Sarita and Jane) and two boys (Tony and Leon). The group is doing work on a unit designed to foster the exploration of ratio situations. They are mixing lemonade concentrates, made of measured spoonfuls of sugar and bottled lemon juice, which they then taste, rate for sweetness, and work with in various ways. The group decides that it needs to make more of its lemonade concentrate, which is a 1:2 mixture (one spoonful of sugar to two spoonfuls of lemon juice). The small quantity they have (one spoon of sugar and two of lemon juice) is not enough for the tasting procedure. But they encounter a dilemma: If they mix another batch of 1:2, which will be one spoonful of sugar and two spoonfuls of lemon juice, and add it to the one they have, they will then have two and four spoonfuls, or a 2:4 mixture. Is this the same mixture?

Jane argues heatedly that two batches of 1:2 would taste the same as one batch of 2:4. The ratios are equivalent, so the mixtures will taste the same. Sarita maintains that they are different mixtures and will not taste the same. Finally they decide to prepare both and conduct a taste test, but Jane says that Sarita cannot be the judge, because she would be "prejudiced," interested only in supporting her own theory. The teacher nominates Tony to be taste-tester. Tony, with a flourish, drinks both mixtures and pronounces the 1:2 mixture and the 2:4 mixture "different. They taste different."

Jane is disappointed, but even Sarita does not accept the judgment as reliable. They both immediately begin to discuss what is potentially wrong with Tony's judgment. They tell me, "Oh, Tony always does this. He just wants to say what Sarita wants to hear. He doesn't know." They begin to discuss ways to expose Tony's unreliability as a taster and as a reporter of his experiences. No energy is put into looking for other sources of error, such as measurement error, a topic the class subsequently spends a great deal of time on, and which is clearly within the grasp of both girls. The search disintegrates into a complaint session about Tony.

In this group, a spontaneous dilemma did lead to some important attempts at sense making. The small group intermental context seemed here to be fostering some authentic "horizontal" group thinking: peers working together to try to figure something out. But for reasons we can only surmise, Tony is not really accepted as a legitimate participant within the context of problem solving, and the part he plays in the interaction itself becomes a focus of anger and conflict.

This leads us to the question of how the classroom must be structured to enculturate students into the habits of mind that let them sustain productive argument that does not disintegrate into personal animosities that overtake the intellectual content. This is pedagogically and theoretically an unsolved problem.

Negotiated Defining and Other Metalinguistic Activities

Most of us view definitions as objects that live within the repository we call the dictionary. But within the view proposed here, they are much more. A *definition* is a sociocognitive entity that occurs throughout many forms of disciplinary intellectual work and practical "everyday" work as well. Community members use informal labels for different kinds of definitions, including *stipulative definitions* (those cases in which speakers agree to assign a particular meaning to a word – a meaning that it may normally not carry – as part of a particular textual or interactional episode); *working definitions* (those which are explicitly part of a developing, exploratory activity); *dictionary definitions* (the conventional and institutionally inscribed community agreements about the meaning of a particular expression); *operational definitions, formal definitions,* and so on. In the work of scientists, social scientists, mathematicians, and others, these types of definitions emerge during joint work activity.

Each kind of definition is a metalinguistic object that entails a set of possible relationships between the actors that use it. When we agree to use a dictionary definition, we have tacitly agreed to take part in the fixed views, beliefs, and values of the community that constructed that definition, and in doing so we may in fact add to its institutionalized power. We are licensed to take part in the world that it indexes. When we agree to the stipulated definition of another intellectual actor, we confer on that person a particularly powerful role. The authoring of a stipulated definition requires a sense of entitlement – a sense of one's own power in a sociointellectual world. An account of stipulative defining that focused merely on the cognitive prerequisites and consequences would, in this view, be seriously incomplete.

What do such activities have to do with children learning mathematics? The joint metalinguistic activity engaged in by working scientists and mathematicians as they construct shared definitions is a signal example of what we mean by authentic intellectual practices of mathematics and science. Dictionary definitions have long been a staple of elementary school math and science, and precise use of such definitions allows students to refer to objects and processes in an unambiguous and technically correct fashion. They are as important as ever, but in reform classrooms, other kinds of defining must take place. In traditional classrooms the creation of one's own definitions was rarely required. But in classrooms that attempt to support "real" inquiry in math and science, the need to agree on the precise meaning of an expression within the local context arises frequently. Within open-ended projects, it is sometimes necessary to develop a working definition of some phenomenon or process, a definition that will change as understanding increases. In such classrooms, the development of one's own symbols, measures, and terminology is often required, with complex and poorly understood difficulties often arising. (See Godfrey & O'Connor, 1995, and discussion below.) Stipulative definitions also arise in the mathematical context when an everyday word carries a restricted meaning within the mathematical domain. Then the ordinary sense of the term should, ideally, be explicitly restricted. In fact, it is fairly easy to imagine instances of these various types of defining within most of the mathematical "modes of cognition" mentioned above: modeling, symbolism, inference, logical analysis, and abstraction.

Within inquiry or reform-type classrooms I have observed a number of instances of a speech activity that allows for the development of such definitions by all members of the group. This speech activity consists of episodes of discussion about word meaning. But these episodes do not follow a script in which the teacher quizzes the students on what a particular word means, nor do they necessarily end with students "looking up the word." Instead, they may contain extended and complex (and usually messy) discussions of word meanings, the complex conditions of their use, and contention over exemplars of their use. I have come to call this speech activity *negotiated defining.* In O'Connor (1992) I made the claim that this kind of activity provided a crucial language socialization experience that bore directly on some aspects of mathematical and scientific modes of thinking and communicating. I will review some of those details below. First I will give some other examples of this phenomenon I have found in the literature.

What Do You Mean by "Even"?

Russell and Corwin (1993) describe changes in classrooms where teachers were trying to create the conditions for mathematical discourse.

One of the changes they find is that teachers have adopted a new attitude toward the definitions of terms used by students. In some cases these teachers became sensitive to occasions where students were using a particular term or expression in "uncomfortable" ways. One fourth-grade teacher stopped a student counting by twos and asked, "What do you mean by *even*?" Here is an excerpt of the transcript Russell and Corwin provide:

Student 1: You add them up, and they add up to even.
Martha [Teacher]: What is even?
Student 2: On one chart you got 1, 3, 5, 7, 9 – and on the other 2, 4, 6, 8, 10.
. . .
Student 4: Odd is if you have three apples. You couldn't split it with a friend.
Martha: Split how? Why can't I get two and you get one?
Student 4: No, it's not even. If we had four, you could have two and I could have two.
Martha: Is 5 even?
. . .
Student 6: If you have five apples, we each could get two and split the other half . . .

(Russell & Corwin, 1993, p. 558)[6]

Notice that the teacher is participating in finding exemplars here, testing the conditions of the term's extension. She does not go straight to the intension, providing a dictionary definition. Students inductively try to derive the conditions that constitute the intension. By coming up with an example, building on the situation introduced by another student, a crucial moment is reached. Russell and Corwin don't say what happened. In order to understand these powerful but intimidating moments, we will need detailed accounts of how they are followed up.

The example does indicate, however, that socialization into a desired "habit of mind" may well be taking place: awareness that others are interpreting and assigning significance to what one is saying. This awareness in and of itself may be a spur to further expressive precision. This is a metalinguistic event, one in which discussion and reflection are welcomed about a previous event of communication. (See also Cobb et al., 1993, for discussion about metalinguistic conversations in mathematics classrooms.)

Points and Corners

In problem-solving situations, students frequently come up with their own terms and words for objects or relations. Often they will find that they are using their informal terms in different ways. The crucible of problem solving, particularly when specific formulations of relationships are called for, will often make these discrepancies apparent in ways that normal interaction would not. What issues arise in such cases? The examples above presented plenty of problems for teachers: when to intervene, when to cite a higher authority, when to call for an end to the discussion. Situations in which students have problems with their own informal terminology may be just as productive intellectually, but they present similar problems for the teacher.

In the following example, the students had somehow started to use the terms *points* and *corners* (Russell and Corwin do not say how this came about). After a time they discover that they do not share a common meaning. In this case too, their use of the terms is partially anchored in everyday use of language and partially extends into the realm of talking about mathematical objects and relations.

> Anna's fifth-grade class was exploring the relationships of the faces, edges, and vertices of various pyramids. As the students compared their results, disagreements emerged about some of the conjectures they were positing. They gradually realized that the disagreement hinged on their definitions of *points* and *corners.* Some students had defined corners to include all the vertices of the pyramid, while others insisted that corners were only the corners of the base and that the vertex on top was "a different kind of thing." (Russell & Corwin, 1993, p. 558)

This teacher judged the exploration and the "roaring discussion" to be productive and did not provide closure at the end of the lesson. Instead, she pointed to the formulas that had emerged from the discussion and said, "Now these are *questionable.* Tomorrow this is what you're going to have to do: decide if your definition really matters" (Russell & Corwin, 1993, p. 558).

Russell and Corwin do not comment on the teacher's final directive, but this seems to be a critical point in classroom communication about such issues. What hinges on the differences in word meanings? Particularly when the words chosen are students' choices, not technical terms with a history, an important question is just that: Why does it matter? There are many answers to this question, as examples below indicate.

Length and Width

It is easy to encounter classroom examples where interpretive conventions are the problem, but the conditions of word use are far more complex than points and corners – perhaps too complex to be decided. In these cases the teacher can be faced with a dilemma. In O'Connor (1992) I described a group discussion that took place over two consecutive fifty-minute class sessions in a sixth-grade classroom in which I had been observing over a period of months. What occurred during these classes was unplanned. The class had been given a measuring task using student-invented measuring units. One object to be measured was a rectangular window, in which the long side of the rectangle extended horizontally and was about twice as long as its up-and-down dimension. A dispute erupted in which some students claimed that the length of the window was its long horizontal dimension, and other students claimed that its length was the up-and-down dimension.

As the discussion ensued, students and teacher adduced example after example of objects from around the room, asking the disputants and their allies what they would label the width and the length. Some of these objects all students agreed upon. Other objects they disagreed about. Finally, two intensional descriptions of the terms *length* and *width* emerged. In one, length was asserted to be the longest side and width was (somewhat more weakly) asserted to be the side perpendicular to the side labeled length. In the other definition, length was asserted to be the dimension of the object that ran parallel with the long axis of the measurer's body. Width was strongly asserted to be "side-to-side."

At first as I observed I was somewhat dismayed at this latter definition. Its proponents even insisted in some cases that the measured length of an object would change if its orientation relative to the observer changed. This seemed truly bizarre to me, but I kept my opinions to myself. I soon found out, by consulting the crosslinguistic literature on spatial language, that virtually all languages offer the option of interpreting spatial terms in one of two ways: One is *deictic,* that is, anchored to the position or orientation of the observer, and the other is *object-oriented,* or "autonomous," that is, computed completely in terms of the internal structure of the entity it is describing. A simple example from English is the following. Imagine yourself standing on a sidewalk, leaning against a storefront, facing the street, with a car parked parallel at the curb. Your friend has gotten out of the car and is looking for her wallet. You say, "Oh, you dropped it, it's in front of the car." There are two places the wallet could be: the deictic interpretation, using you the speaker as the anchor, would lead to the conclusion that the wallet is between you and the passenger side of the car in front of you. The object-centered interpretation would lead to the conclusion that the wallet is somewhere around the headlights, in front of the canonical "front" of the car: When the car is facing forward, its front is where the headlights are.

The students split along these two definitions not by chance, but because there is a pervasive opposition between deictic and autonomous uses of spatial terms in language in general. Now, that is not to say that the two definitions are mathematically equivalent. In fact, as Pimm (1987) notes, the mathematical register relies completely on object-oriented, autonomous uses of spatial terms. We cannot have the altitude of a triangle changing just because I orient the page differently toward myself.

So these students and teacher had fallen into a set of usage conditions with several layers. It was hard to put one's finger on just what should be done, since both groups felt strongly about their definitions. They were standing at the boundary between everyday uses of language and technical or "mathematics register" uses, and the tension could be felt. Some of those who supported the autonomous definition became distant and disgusted, and a few of those who supported the deictic definition became almost belligerent.

When the dictionary was finally consulted, it was no help. It gave a definition of *length* as "the measurement or distance of the longest side," an autonomous-style definition, but it defined *width* as "the distance from side to side," a distinctly deictic view! Each group claimed victory and began arguing again.

Edwards (1989), discussing Walkerdine's views on school mathematics practices, puts this well: "Discursive practices at school, while often aimed at introducing formal concepts within friendly and familiar contexts, can easily result instead in 'a complex and bewildering confusion' . . . we get . . . a clash between two discourses, with different significations – same signifiers, different signifieds, so different signs" (p. 47).

Note that unlike some of the episodes of defining I recounted earlier, this episode featured two words – *length* and *width* – that are part of everyone's everyday inventory. As such, all speakers generally feel qualified to rule on their correct use and do not take kindly to the suggestion that their interpretation is "wrong." In this case, however, one set of meanings was clearly more aligned with the expressive norms of mathematics: the meanings that were object-centered and nondeictic.

Perhaps the most problematic issue to arise in such discussions is one of authority. In general, people believe there is one correct meaning for most words. In the midst of a discussion in which students come up with quite different interpretations for a word, it is common to observe disgust, anger, even outrage. Teachers also may feel a deep sense of confusion upon entertaining discussion about new uses of a word they feel they know well. Students will turn to them for reassurance about word meanings.

The Vertical Handspan

I will present one more example of an extended episode centering on word meaning, reported on at length in Godfrey and O'Connor (1995). This example is intended to give readers a sense of the wide range of discipline-based values and practices encountered in these episodes of negotiated defining: These interactions are often far from simple. Here, a quasi-constructivist instructional activity called for sixth-grade students to create their own units of measurement. The students were to present their units and a notation they had crafted for creating sentences comparing the heights of two individuals. One boy used his hand, from his wrist to the tip of his middle finger. This he named "a Kadeem handspan." Another student in the class, Elliot, and several others began to describe what they thought the word *handspan* meant, showing a spread-fingered hand and indicating the width of their hands. Elliot said "handspan is like wingspan – from tip to tip." Elliot told Kadeem he couldn't do what he had done – he could not make up a new meaning for a word like handspan. Kadeem then turned to the teacher and said *he* had thought it was OK in this class to create new meanings for words. This moment held tensions that are felt far beyond that particular sixth grade. Issues of tradition, authority, legitimacy, and intelligibility are all at stake.

Other students suggested changing the name to *hand length* or even *vertical handspan*. Elliot continued to insist they not use the word handspan at all. The teacher asked him to explain why he felt so strongly about this. He replied, "Because everybody knows that when you say handspan you mean across the hand, the width, and if they use it to mean the length, people will get really confused and nobody will understand their symbols." The teacher summed up that day by pointing to two results: There is a conventional meaning of handspan, and people had suggested a modification, vertical handspan. The next day, Kadeem's group had decided to change the name to vertical handspan.

The interfacing of the standard definition and the student's creation is an issue inherent in the very nature of the curricular process and the methods used in this curriculum. Other classrooms where student invention and construction of units and symbols are considered useful and productive activities no doubt encounter similar conflicts. And these are conflicts with no easy resolution. Elliot has linguistic authority on his side in the form of a codification of the history of English – the dictionary. Moreover, Kadeem needs to be cognizant of the conventional, community-based nature of word meaning. On the other hand, we can be fairly certain that Kadeem knows the power of conventional word meaning. Every day, he and all the other students in the class use thousands of words, each of which has conventional meanings, meanings Kadeem would not think of expanding in a nonstandard direction. Here the activity of negotiated defining evoked the tensions between authority and innovation that are played out in the most rarefied academic contexts – often to no clear resolution.

Protoforms of Negotiated Defining

We have all witnessed children arguing about the meaning of a word, whether this takes place in the context of a game, where the definition will determine winners and losers, or in the context of verbal play, or in conflict for its own sake. In these early episodes, the central themes of authority, collaboration, and entitlement emerge clearly. Who gets to decide the meaning or significance or operational definition of a word or expression? What will it take to get one's peers to agree to a stipulated definition? What hinges on the dispute?

The same questions arise in classroom episodes of negotiated defining, but in addition, given the right sort of support, the following understandings may result, all of which are an important part of the discourse practices of mathematical work:

- knowledge that one is accountable for the meaning of the words one uses
- knowledge that different contexts can license different interpretations of the (superficially) same word or expression
- understanding that choice of use of linguistic expression within a context is part of human action and the expression of intention to a community of interlocutors
- knowledge that for a new context one can decide upon a signifier–signified pairing that has agreed-upon properties
- knowledge that one can stipulate a meaning, if one's colleagues will allow it

These pieces of knowledge and belief are prerequisites to the kind of metalinguistic flexibility needed to engage collaboratively in the mathematical modes of cognition cited above. To those who have mastered these principles and beliefs they may seem self-evident and transparent: What's to learn? Yet this is a characteristic of all kinds of complex socialization: Once one has become fully socialized, there seems to be no other way; and before one is fully competent, the way ahead seems opaque and untraversable. It is clear that if we wish to understand the relationships among home-based protoforms of mathematical practices, the practices of the school mathematics, and those of the professional discipline, we must gain a much deeper understanding of the workings of all of these discourse practices.

Summary

Through all of these examples of protoforms and advanced forms of mathematical and scientific discourse, two questions recur, unresolved. One involves the availability of early discourse protoforms to students' in-school appropriation of mathematical reasoning. The "transfer" of discourse expertise will not be easy in many cases, for what are no doubt complex amalgams of mathematical, social, and experiential factors. One important factor involves access to different discourse genres: Whether inside or outside the classroom, who is entitled to argue, to speculate, to define? Each of these speech activities involves some relation to what Penelope Eckert refers to as "meaning-making rights" (Eckert, in press), and thus they will be only partially available to a variety of people in a variety of contexts, as determined by a whole host of social factors. Why might some children be entitled to make meaning on the playground and yet not participate in the classroom, or vice versa? This is a larger question than those posed in this chapter, yet it bears an important relationship to their eventual answers.

The second question is pedagogical: Given the difficulties discussed above, how can teachers create the conditions that foster the expansion of students' discourse repertoires into mathematical and scientific domains? This is a question that Ochs et al. raise in their work on protoforms of theorizing:

> We propose that complex storytelling in which perspectives are challenged and redrafted collectively is more likely to occur where co-narrators are familiar with one another and/or the narrative events than where co-narrators are socially distant. . . . Indeed, elementary and secondary schools may face a severe handicap in trying to instill scholarly processes of theory-building, critiquing, and redrafting if the environment of instruction lacks the familiarity that is characteristic of more intimate family or even university research settings. While Vygotsky attributed to schools a vital role in socializing "scientific concepts"

and the processes of evaluation, critique and redrafting (Vygotsky, 1986), our present schools may have lost the familiarity necessary for students to undertake meaningful engagement in these processes. In other words, size of the co-narrating and co-cognizing group may be inversely related to the facilitation of theory-building and deconstructing. (Ochs et al., 1992, p. 67)

With that statement, Ochs opens up a crucial and volatile set of issues about community and socialization that are, at least for the present, beyond the scope of this work.

Acknowledgments

I am grateful to the colleagues who have commented on the material in this chapter, including Lynne Godfrey, Annabel Greenhill, Magdalene Lampert, Sarah Michaels, Pamela Paternoster, and all the participants at the 1994 Conference on Classroom Communication, sponsored by the National Center for Research on Mathematics and Science Education, University of Wisconsin, Madison. All remaining errors are mine.

Notes

1. As will become clear below, I am not claiming, as do some discourse theorists, that there is no cognition beyond discourse practices. Rather, I agree with the many sociocultural theorists who have argued that the development of "cognitive abilities" must be understood in relation to the social practices within which learners first encounter and practice them.
2. Some readers will note that I have obscured an important distinction here. Am I saying that mathematical practice, and thus mathematical knowledge, is purely social, in opposition to thousands of years of mainstream philosophy (but in accord with a small minority of mathematicians and sociologists, struggling for "a new understanding of mathematics . . . grounded in social realities rather than metaphysical and psychological fictions" [Restivo et al., 1993, p. 9])? Or am I simply saying that mathematical knowledge and its deployment are most usefully thought of, pedagogically speaking, as being carried out in the context of joint action, which almost invariably involves many kinds of discourse? Here I will only say that I intend the points I make to be intelligible and perhaps useful to those at both ends of this quasi-continuum. See also note 4.
3. Obviously, this does not mean that any classroom discourse practice results in the kind of learning that reform-minded mathematics educators are seeking. Some discourse practices instantiate a socialization process that results in students who despise mathematics and science, or who have only the most tenuous grasp of its power and complexity. In this chapter I will be engaging in something of an idealization for the purpose of exploring classroom discourse contexts as productive sites for socialization into some forms of mathematical thinking. I do not intend to imply that such contexts will be easy to create or to evaluate.

4. This view is elaborated upon in several different traditions. For example, sociocultural theorists (Cole & Scribner, 1975; Scribner & Cole, 1981; Vygotsky, 1978, 1986; Wertsch, 1985, 1991) would assert that ways of thinking such as these are organized, constituted, and maintained at least partially through the situated enactment of multiparty activities of various kinds. Practitioners of an ethnomethodological sociology of science would assert that the goals, values, and practices that constitute the discourse of a field or discipline are often embodied in routines or forms of interaction that result in the production of particular kinds of texts, both oral and written (Knorr-Cetina, 1981; Latour & Woolgar, 1986; Lynch, 1985).

   There are numerous different positions on the relationship between modes of cognition and social practices in school-based learning. Edwards (1993) takes the extreme view that the analysis of classroom teaching and learning requires no explanatory appeal to concepts, memories, or other cognitive states; the analyst need not go beyond language-mediated discursive practices (Edwards, 1993, p. 220) – discourse *is* the object of cognitive study. Pontecorvo and Girardet (1993) express a more moderate view. Pontecorvo (1993, p. 191) states that "forms of discourse *become* forms of thinking. Indeed, thinking methodologies of the specific domain are enacted through appropriated discourse practices that respond to the epistemic needs of a disciplinary topic." In this view, "speech is not considered a tool that guides or indirectly points to some 'material' actions. Rather, 'what has to be done' together is a discursive action, and the action that has to be carried out through discourse is a social knowledge construction that is the object of ongoing negotiation between participants" (Pontecorvo & Girardet, 1993, p. 366) [emphasis mine].
5. Walkerdine (1988) discusses the home-based social practices and ideologies that even preschoolers have solidly installed and convincingly shows that these cannot be factored out or ignored in favor of an idealization that views school-based activities as autonomous. Lave and Wenger (1991) and Lave (1988) develop similar insights about adults. For the purposes of this chapter, however, I will emphasize classroom discourse often to the exclusion of home discourse, in full awareness that this picture is seriously incomplete.
6. In this cited transcript, some numbers are written as numerals and others are spelled out as they were in the original text. In transcripts original to this volume, we adhere to the convention of spelling out numerals in dialogue.

References

American Association for the Advancement of Science. (1993). *Benchmarks for science literacy.* New York: Oxford University Press.

Bauman, R., & Scherzer, J. (Eds.). (1989). *Explorations in the ethnography of speaking* (2nd ed.). New York: Cambridge University Press.

Bearison, D. J., Magzamen, S., & Filardo, E. K. (1986). Socio-conflict and cognitive growth in young children. *Merrill-Palmer Quarterly, 32,* 51–72.

Cobb, P., Wood., T., & Yackel, E. (1993). Discourse, mathematical thinking, and classroom practice. In E. Forman, N. Minick, & C. A. Stone (Eds.), *Contexts for learning: Sociocultural dynamics in children's development,* pp. 91–119. Oxford: Oxford University Press.

Cole, M., & Scribner, S. (1975). Theorizing about the socialization of cognition. In T. Schwartz (Ed.), *Socialization as cultural communication,* pp. 157–176. Berkeley: University of California Press.

Eckert, P. (in press). Discussant remarks. In J. Greeno & S. Goldman (Eds.), *Thinking practices.* Hillsdale, NJ: Erlbaum.

Edwards, D. (1989). The pleasure and pain of mathematics: On Walkerdine's *The mastery of reason. The Quarterly Newsletter of the Laboratory for Comparative Human Cognition, 11* (1–2), 38–41.

Edwards, D. (1993). But what do children really think? Discourse analysis and conceptual content in children's talk. *Cognition and Instruction, 11* (3–4), 207–225.

Eisenberg, A., & Garvey, C. (1981). Children's use of verbal strategies in resolving conflicts. *Discourse Processes, 4,* 149–170.

Forman, E., & Cazden, C. (1985). Exploring Vygotskian perspectives in education: The cognitive value of peer interaction. In J. V. Wertsch (Ed.), *Culture, communication and cognition,* pp. 323–347. New York: Cambridge University Press.

Godfrey, L., & O'Connor, M. C. (1995). The vertical handspan: Nonstandard units, expressions, and symbols in the classroom. *Journal of Mathematical Behavior, 14,* 327–345.

Goodwin, M. H. (1990). *He-said-she-said: Talk as social organization among black children.* Bloomington, IN: Indiana University Press.

Gray, L. (1993). *Large group discussion in a 3rd/4th grade classroom: A sociolinguistic case study.* Unpublished doctoral dissertation, Boston University, Boston, MA.

Greeno, J., & Goldman, S. (Eds.). (in press). *Thinking practices.* Hillsdale, NJ: Erlbaum.

Gumperz, J. J., & Hymes, D. (1972). *Directions in sociolinguistics: The ethnography of communication.* New York: Holt, Rinehart & Winston.

Hatano, G. (1993). Time to merge Vygotskian and constructivist conceptions of knowledge acquisition. In E. A. Forman, N. Minick, & C. A. Stone (Eds.), *Contexts for learning: Sociocultural dynamics in children's development,* pp. 153–166. Oxford: Oxford University Press.

Hatano, G., & Inagaki, K. (1991). Sharing cognition through collective comprehension activity. In L. Resnick, R. Levine, & S. Teasley (Eds.), *Perspectives on socially shared cognition,* pp. 331–348. Washington, DC: American Psychological Association.

Heath, S. B. (1983). *Ways with words: Language, life, and work in communities and classrooms.* New York: Cambridge University Press.

Hiebert, J., & Wearnes, D. (1993). Instructional tasks, classroom discourse, and students' learning in second-grade arithmetic. *American Educational Research Journal, 30* (2), 393–425.

Hymes, D. (1962). The ethnography of speaking. In T. Gladwin & W. C. Sturtevant (Eds.), *Anthropology and human behavior.* Washington, DC: Anthropological Society of Washington. [Reprinted in J. A. Fishman, *Readings in the sociology of language.* 1968. The Hague: Mouton.]

Hymes, D. (1974). Ways of speaking. In R. Bauman & J. Sherzer (Eds.), *Explorations in the ethnography of speaking.* London: Cambridge University Press.

Hymes, D. (1981). *In vain I tried to tell you: Essays in Native American ethnopoetics.* Philadelphia, PA: University of Pennsylvania Press.

Knorr-Cetina, K. (1981). *The manufacture of knowledge.* Oxford: Pergamon Press.

Lampert, M. (1990, Sept. 3). *Practices and problems in teaching authentic mathematics in school.* Plenary address presented at the international symposium "Research on effective and responsible teaching," Fribourg, Switzerland.

Lampert, M., Rittenhouse, P., & Crumbaugh, C. (1996). Agreeing to disagree: Developing sociable mathematical discourse in school. In D. R. Olson & N. Torrance (Eds.), *Handbook of psychology and education: New models of learning, teaching, and school,* pp. 731–764. Oxford: Basil Blackwell.

Latour, B., & Woolgar, S. (1986). *Laboratory life* (2nd ed.). Princeton, NJ: Princeton University Press.

Lave, J. (1988). *Cognition in practice.* London: Cambridge University Press.

Lave, J., & Wenger, E. (1991). *Situated learning: Legitimate peripheral participation.* New York: Cambridge University Press.

Leer, M., Bill, V., & Resnick, L. (1993, April). *Mathematical power and responsibility for reasoning: Forms of discourse in elementary mathematics.* Paper presented at the Annual Meeting of the American Educational Research Association, Atlanta, GA.

Lynch, M. (1985). *Art and artifact in laboratory science.* London: Routledge & Kegan Paul.

Michaels, S. (1982). "Sharing time": Children's narrative styles and differential access to literacy. *Language in Society, 10,* 423–442.

National Council of Teachers of Mathematics. (1989). *Curriculum evaluation standards for school mathematics.* Reston, VA: National Council of Teachers of Mathematics.

National Research Council. (1989). *Everybody counts: A report to the nation on the future of mathematics education.* Washington, DC: National Academy Press.

O'Connor, M. C. (1992). *Negotiated defining: The case of length and width.* Unpublished manuscript, Boston University.

O'Connor, M. C. (1996). Managing the intermental: Classroom group discussion and the social context of learning. In D. Slobin, J. Gerhardt, A. Kyratzis, & J. Guo (Eds.), *Social interaction, social context, and language,* pp. 495–509. Hillsdale, NJ: Erlbaum.

O'Connor, M. C., & Michaels, S. (1996). Shifting participant frameworks: Orchestrating thinking practices in group discussion. In D. Hicks (Ed.), *Discourse, learning and schooling,* pp. 63–103. New York: Cambridge University Press.

Ochs, E. (1991). Indexicality and socialization. In J. W. Stigler, R. A. Shweder, & G. Herdt (Eds.), *Cultural psychology: Essays on comparative human development,* pp. 287–308. New York: Cambridge University Press.

Ochs, E., Taylor, C., Rudolph, D., & Smith, R. (1992). Storytelling as a theory-building activity. *Discourse Processes, 15,* 37–72.

Orsolini, M., & Pontecorvo, C. (1992). Children's talk in classroom discussions. *Cognition and Instruction, 9,* 113–136.

Pimm, D. (1987). *Speaking mathematically: Communication in mathematics classrooms.* London: Routledge & Kegan Paul.

Pontecorvo, C., & Girardet, H. (1993). Arguing and reasoning in understanding historical topics. *Cognition and Instruction, 11* (3–4), 365–395.

Putnam, R. T., Lampert, M., & Peterson, P. (1990). Alternative perspectives on knowing mathematics in elementary schools. In C. Cazden (Ed.), *Review of Research in Education* (Vol. 16, pp. 57–144). Washington, DC: American Educational Research Association.

Resnick, L., Levine, R., & Teasley, S. (Eds.). (1991). *Perspectives on socially shared cognition.* Washington, DC: APA.

Restivo, S., Van Bendegem, J. P., & Fischer, R. (Eds.). (1993). *Math worlds: Philosophical and social studies of mathematics and mathematics education.* Albany, NY: State University of New York Press.

Russell, S. J., & Corwin, R. B. (1993, March). Talking mathematics: "Going slow" and "letting go." *Phi Delta Kappan,* 555–558.

Sanches, M., & Blount, B. (Eds.). (1975). *Sociocultural dimensions of language use.* New York: Academic Press.

Schiffrin, D. (1985). Everyday argument: The organization of diversity in talk. *The handbook of discourse analysis: Vol. 3. Discourse and Dialogue,* pp. 35–46. London: Academic Press.

Schoenfeld, A. (1989). Problem solving in context(s). In R. Charles & E. Silver (Eds.), *The teaching and assessing of mathematical problem solving,* pp. 82–92. Reston, VA: NCTM.

Scollon, R., & Scollon, S. (1981). *Narrative, literacy, and face in interethnic communication.* Norwood, NJ: Ablex.

Scribner, S., & Cole, M. (1981). *The psychology of literacy.* Cambridge, MA: Harvard University Press.

Swing, S. R., & Peterson, P. L. (1982). The relationship of student ability and small-group interaction to student achievement. *American Educational Research Journal, 19,* 259–274.

Toulmin, S. E. (1958). *The uses of argument.* London: Cambridge University Press.

Vygotsky, L. (1978). *Mind in society: The development of higher psychological processes.* M. Cole, V. John-Steiner, S. Scribner, & E. Souberman (Eds. and Trans.). Cambridge, MA: Harvard University Press.

Vygotsky, L. (1986). *Thought and language.* A. Kozulin (Ed. and Trans.). Cambridge, MA: MIT Press. [Original work published 1962.]

Walkerdine, V. (1988). *The mastery of reason: Cognitive development and the production of rationality.* London: Routledge.

Webb, N. M. (1991). Task-related verbal interaction and mathematics learning in small groups. *Journal for Research in Mathematics Education, 22* (5), 366–389.

Wertsch, J. V. (1985). *Vygotsky and the social formation of mind.* Cambridge, MA: Harvard University Press.

Wertsch, J. (1991). *Voices of the mind: A sociocultural approach to mediated action.* Cambridge, MA: Harvard University Press.

Wertsch, J. V., & Rupert, L. (1993). The authority of cultural tools in the sociocultural approach to mediated agency. *Cognition and Instruction, 11,* 227–240.

Yackel, E., Cobb, P., & Wood, T. (1991). Small group interactions as a source of learning opportunities in second-grade mathematics. *Journal for Research in Mathematics Education, 22,* 390–408.

3 The Role of Imagery and Discourse in Supporting Students' Mathematical Development

Kay McClain and Paul Cobb

The current reform movement in mathematics education places particular emphasis on the role of classroom discourse in supporting students' mathematical development. In particular, the *Professional Standards for Teaching Mathematics* (National Council of Teachers of Mathematics [NCTM], 1991) argue that "the nature of classroom discourse is a major influence on what students learn about mathematics" (p. 45). The teacher's role in guiding the development of productive mathematical discourse includes deciding when to allow students to struggle with an idea, when to ask questions, and when to tell. These decisions are based on the teacher's judgments about the ways individual students might be able to reorganize their thinking as they participate in communal classroom processes. In this manner, teachers can be seen to orchestrate classroom discourse based on their understandings of the mathematics they are teaching and of their students' reasoning (NCTM, 1991).

In this chapter, we focus on the relationship between discourse as a communal or collective activity and the mathematical development of students as they participate in it and contribute to its development. Our particular concern is with a process that we call the *folding back* of discourse, wherein the mathematical relationships under discussion are redescribed in terms of the specific situations from which they emerged. We conjecture that the participation of individual students in this collective process constitutes a supportive context for them to ground their increasingly sophisticated mathematical activity in situation-specific imagery. This, for us, is an important aspect of mathematics instruction that has understanding as its first priority.

In the first part of the chapter, we discuss the notions of imagery and discourse and the role that students' participation in discourse can play in enabling them to ground their activity. We develop these issues further in the remainder of the chapter by drawing on data collected during a yearlong teaching experiment conducted in a first-grade classroom.

To provide an orientation, we describe the classroom setting and the instructional theory that guided the development of instructional sequences during the experiment. Against this background, we then analyze three classroom episodes to clarify issues relating to both the folding back of discourse and the mathematical activity of students as they participate in it.

Imagery and Discourse

Thompson (in press), speaking in psychological terms, argues that meaningful "mathematical reasoning at all levels is firmly grounded in imagery." He supports this contention by discussing various forms of imagery which include personal images that inform teachers' and students' mathematical activities, teachers' and students' background images, images of notational activity, and imagery of situations. In doing so, he draws our attention to the ubiquitous role of imagery in mathematics teaching and learning. We draw on Thompson's work in this chapter but limit our focus to situation-specific imagery that students evoke to support their quantitative reasoning. As Thompson (in press) notes, difficulties often arise for students because of insufficient attention being given to their images of the settings in which problems ostensibly occur. This often results in students' activity becoming "decoupled" from their interpretations of problem situations. To clarify this point, Thompson discusses students' solutions to the following problem: *Here is a machine [referring to the table shown in Figure 3.1]. Eight pieces of meat are going in and six packages of meat are coming out.* Students are asked to complete the table by finding the outputs when specified amounts of meat go into the machine.

Thompson reports that when students solved this task, most merely looked for patterns in the pairs of numbers in the table and approached the problem as a "guess my rule" activity. "Their conversations were empty in regard to the machine or why the number pairs were related in a natural way" (Thompson, in press). Thompson argues that, for students, such tasks involve deciphering the relationship between conventional notation schemes and the "superficial characteristics of a problem statement's linguistic presentation" (pp. 15–16). To encourage students' development of imagery of the situation in the problem statement, the meat packing task was subsequently changed to a scenario about a machine that turns seven round chunks of bologna into eight rectangular packets. As a consequence of this modification, the students discussed the machine and the process of cutting up the original pieces of bologna to

Amount of meat entered into the machine	Number of packages resulting
8	6
7	
5	

Figure 3.1. Table for bologna-packing problem

repackage them. This led to conversations about cutting up an amount of bologna into sevenths and then repackaging the same amount into eighths. Thompson's analysis indicates that the students' construction of imagery of the scenario helped to initiate a shift in their discussion so that they were talking about their interpretations of the problem situation instead of simply generating a table of numbers that did not signify quantitative relationships in that situation.

Pirie and Kieren's (1989) recursive theory of mathematical understanding substantiates Thompson's arguments about the important role that imagery plays in supporting students' development of mathematical understanding. Pirie and Kieren contend that mathematical understanding is a "recursive phenomenon and recursion is seen to occur when thinking moves between levels of sophistication" (p. 8). Initially, students make images of either their situation-specific activity or its results. The first significant development occurs when students can take such images as givens and do not have to create them anew each time. This development is supported by students' engagement in instructional tasks in which they anticipate acting and subsequently reflect on its consequences. Mathematical understanding continues to develop as students reason with their images, initially noticing properties and subsequently taking these properties of images as givens that can be formalized. One of the strengths of Pirie and Kieren's analysis of the growth of mathematical understanding is its broad scope in tracing development from the creation of images to formalization and axiomatization.

It is important to note that, for Pirie and Kieren (1989), the actual process of learning does not proceed in a linear manner through a sequence of levels. While activity at a more concrete inner level of knowing can serve as a starting point from which to build more sophisticated mathematical ways of knowing, students often *fold back* to activity at a previous level to provide grounding for their subsequent mathematical activity. In this manner, the students' imagery continues to support development in a recursive process, such that increasingly sophisticated mathematical activity is informed by interpretations of specific situations. When students fold back to a previous level, however, the action at that prior level is not the same. The experiences at the outer level serve to inform the revisited inner level. This notion of folding back is rooted in imagery and consistent with Thompson's analysis. We should stress, however, that Pirie and Kieren's notion of folding back is cast in psychological terms and was developed to account for shifts in individual activity. We will complement this focus by speaking of the folding back of discourse when we discuss communal classroom processes.

The approach of analyzing the development of both classroom discourse and the mathematical understanding of students who participate in it is illustrated by Cobb, Boufi, McClain, and Whitenack (1997). Cobb et al. describe shifts in collective activity that occur during a type of classroom discourse that they call *reflective discourse.* Reflective discourse is characterized by a shift in the level of discourse such that what is said and done in activity subsequently becomes an explicit object of discussion. Cobb et al. conjecture that students' participation in reflective discourse supports the participating students' progressive mathematization of activity by giving rise to opportunities for them to reflect on and objectify prior activity. In this view, participation in reflective discourse is seen to both support and enable individual mathematization. We should stress, however, that participation in reflective discourse constitutes conditions for the possibility of mathematical learning, but that it does not inevitably result in each student reorganizing his or her mathematical activity.

It is important to note that this account of the supportive role of reflective discourse is based on a psychological perspective in which mathematical understanding is "characterized by the creation and conceptual manipulation of experientially-real mathematical objects" (Cobb et al., 1997). This psychological orientation is consistent with the viewpoints developed by Thompson (in press) and by Pirie and Kieren (1989). It should, however, be differentiated from mainstream American approaches that account for the understanding in terms of the construction of internal representations. In place of information and representation, this perspective has socially situated activity as its basic currency.

Returning to Pirie and Kieren's analysis of mathematical understanding, we can view reflective discourse as the collection or communal analog of a movement in individual students' activity to increasingly sophisticated, outer levels of mathematical understanding. Extending the analogy, the folding back of discourse can be viewed as the communal correlate of the psychological process of folding back to inner levels of mathematical knowing. It therefore involves a recursion of discourse. We develop this notion of the folding back of discourse by presenting three classroom episodes that allow us to explore issues related to imagery and discourse. The episodes should be of more than local interest in that they can serve as paradigmatic cases that can both help teachers develop understandings of their own practice and contribute to the growing research literature on reform teaching.

The Setting

The three episodes occurred in a first-grade classroom in which we conducted a yearlong teaching experiment. As will become clear, the teacher with whom we collaborated participated as a full member of the research and development team.[1] In the following paragraphs, we describe the classroom setting and give an overview of the instructional theory that guided the development of the instructional sequences during the experiment.

Ms. Smith's Classroom

Ms. Smith's class consisted of 11 boys and 7 girls and was one of three first-grade classes in an elementary school in a large suburban area. The majority of the students were from middle- to upper-middle-class families. Most parents had professional occupations. There were no minority students in the classroom, although a small percentage attended the school. The students in the class were representative of the school's general student population. Although not a Christian school, morals and values were part of the responsibility of schooling and students regularly participated in spiritual activities.

Instructional Sequences

The instructional sequences enacted in the project classroom were developed in collaboration with Koeno Gravemeijer of the Freudenthal Institute. The sequences were designed to address the core quantitative concepts while supporting Ms. Smith's efforts to establish an inquiry mathematics microculture in her classroom. Although general phases of the instructional sequences had been outlined prior to the teaching experiment, modifications were made on a daily basis as we conducted initial analyses of individual and collective activity. In a very real sense, the sequences emerged during the experiment as we attempted to make sense of what was happening in the classroom. The interpretive framework that guided this analysis is discussed in Cobb and Yackel (in press).

The instructional theory developed by researchers at the Freudenthal Institute is known as the theory of Realistic Mathematics Education (RME). One of its central tenets is that the starting point of an instructional sequence should be *experientially real* so that students can evoke the imagery of the situations described in problem statements when solving tasks. In this way, the students' construction of situation-specific imagery allows them to engage in personally meaningful mathematical activity and constitutes a basis for the students' subsequent mathematization of activity (Gravemeijer, 1990; Streefland, 1991). Further, it provides a grounding to which they can later fold back to reestablish an imaginistic basis for their activity. In this regard, Thompson (1992) notes that "if students do not become engaged imaginistically in the ways that relate mathematical reasoning to principled experience, then we have little reason to believe that they will come to see their worlds outside of school as in any way mathematical" (p. 10). This tenet of RME can also be seen to be consistent with the recommendations derived from investigations that have compared and contrasted mathematical activity in the classroom with that in out-of-school situations (Nunes, Schliemann, & Carraher, 1993; Saxe, 1991).

As a point of clarification, it should be stressed that the term *experientially real* means only that the starting points should be experienced as real by the students, not that they should necessarily involve everyday situations. Further, we take it as self-evident that even when everyday scenarios are used as starting points, they necessarily differ from the situations as students might experience them out of school (Lave, 1993; Walkerdine, 1988). We therefore see no reason to engage in "everyday" mathematics for its own sake. Our central point is instead that the interactive constitution of the scenario in the classroom with the teacher's guidance contributes to the students' construction of imagery of the situation, and this can provide a grounding for their subsequent mathematical activities.

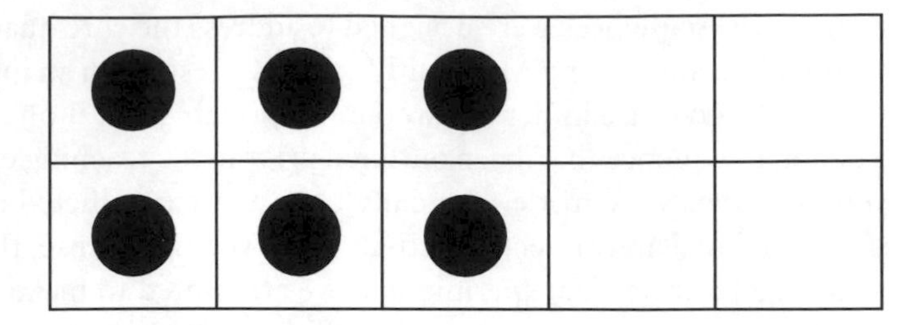

Figure 3.2. Ten frame with 6 chips

Classroom Episodes

An important aspect of Ms. Smith's teaching practice involved guiding the development of a collective narrative related to situations described in instructional tasks. She would typically spend several minutes with the students elaborating a story that could then serve as the basis for problem-solving situations. The students' participation in these narratives about real or fictitious situations supported their construction of imagery of the situation, thus making the described situation experientially real for them. For example, when introducing instructional activities that involved the use of a ten frame (see Figure 3.2), Ms. Smith talked with the students about Earl, a pumpkin seller who was familiar to the students, and explained that his pumpkins were packed in crates of ten. Ms. Smith supported the students' participation in the narrative by asking them to share experiences of buying pumpkins and visiting Earl's pumpkin stand. She then negotiated with the students that a ten frame shown on an overhead projector signified a pumpkin crate and that chips signified pumpkins in the crate (see Figure 3.2).

As a further aspect of her practice, Ms. Smith guided shifts in the level of discourse as students explained their interpretations and solutions. These shifts in the level of discourse frequently made it possible for students' contributions to become objects of collective reflection thereby providing opportunities for students to mathematize their prior activity (Cobb et al., 1997). However, as Ms. Smith initiated and supported these shifts, it sometimes appeared that students' interpretations of the tasks were not always grounded in situation-specific imagery. This was indicated by calculational explanations that did not seem to carry the significance of acting on quantities in the task situation. Ongoing discussions in the research team about these observations led Ms. Smith to reflect on particular classroom episodes and to develop an appreciation for the importance of supporting the grounding of students' activity in imagery.

In each of the episodes that we will present, students' activity appears to lose its initial grounding. In the subsequent classroom discussions, Ms. Smith attempted to provide opportunities for students to interpret ongoing activity in terms of situation-specific imagery. In the first episode, we discuss and clarify the notion of *folding back*. In the second episode, we introduce the notion of *dropping back* and differentiate between it and folding back. In the third episode, we clarify problems in communication that arise when a taken-as-shared basis in imagery has not been established.

The Folding Back of Discourse

The first sample episode occurred in early November and involved an instructional task designed to support students' development of doubles-based thinking strategies (e.g., using $6+6=12$ to find $6+8$). At this particular point in the school year, some but not all of the students routinely used this type of strategy. Further, it appeared that an appreciable number of students did not know certain doubles relationships (e.g., $7+7=14$, $16-8=8$). In collaboration with Ms. Smith, the research team therefore proposed instructional activities in which each of two equal numerical quantities was repeatedly increased or decreased by one (e.g., $10+10$, $9+9$, $8+8$). The first task of this type involved a scenario in which a set of twins collected stickers. Our instructional intent was that the students might generate successive doubles relationships by reasoning about changes in the number of stickers that each twin had. Ms. Smith introduced the scenario as a narrative on November 8 by asking if they were familiar with twins. One of the students, Jane, had twin brothers and was asked to share things about twins. The dialogue between Ms. Smith and the students continued for several minutes so that the scenario might become experientially real for the students. Ms. Smith then posed the following task:

T: Think about twins. It can be twins you know or ones that are just made up in your mind . . . and both of these twins collect stickers. Each one, they collect stickers. They try to keep it even. They keep it so they both have the same number of stickers. Then they want to know how many they have altogether. Now you think about this. If we had two twins and they each had four stickers (shows four fingers on each hand), then how many stickers do they have altogether? Each twin has four stickers, how many do they have altogether? Sue?

Although Ms. Smith attempted to support the students' construction of imagery of the situation, their activity in the subsequent discussion appeared to be purely calculational and to lack grounding. Even when Ms. Smith cast her questions in terms of the twins and their stickers, students' responses were typically purely numerical and did not necessarily signify quantities of stickers either for themselves or for other members of the class. As a consequence, the task as it became constituted in the classroom did not support the generation of the doubles as quantitative relationships. This occurred in part because the classroom discourse focused on calculational solutions rather than task interpretations, making it possible for students to participate effectively by creating patterns in numerals per se.

T: If each twin has four stickers then they have eight altogether. What if each twin has five stickers . . . each twin has five stickers . . . Amy?
Amy: Uhm . . . ten.
T: If each twin has six stickers, each twin has six, Bob?
Bob: Twelve.
T: If each twin has six stickers, then six stickers and six more stickers is twelve. What if each twin has seven stickers, seven stickers then how many . . . Kitty?
Kitty: Fourteen.
T: How did you figure it out?
Kitty: I just know.

As the exchange progressed, Ms. Smith began to ask students to explain their answers. Although Kitty merely stated "I just know," other students gave explanations that supported a focus on numeral patterns. For example, Bob offered the following explanation for seven stickers and seven stickers.

T: Okay, what if . . . (Bob interrupts)
Bob: All you've got to do is count two more from the last one . . . it was twelve so you just added two more to make seven so you just counted two more.

Bob's comment that "you just added two more to make seven" indicates that his explanation may have had quantitative significance for him. However, many students seemed to interpret it as a directive to simply add two to the previous result.

As the exchange continued, two students unexpectedly gave explanations that involved multiplication. First, Dave shared his knowledge of the multiplication facts and then Jane offered another multiplication fact that appeared unrelated to the task at hand.

Dave: I figured it out this way – there were sixteen 'cause I know the multiplication table because two times eight would be sixteen.

T: Raise your hand if you were also thinking that eight stickers and eight stickers would be sixteen. What if each twin had nine stickers? Amy?

Amy: Nineteen 'cause if you had nine and nine more I think it would make nineteen.

T: Jane, what do you think?

Jane: Eighteen.

T: Why do you think it's eighteen?

Jane: 'Cause nine times nine equals eighty-one and nine plus nine is eighteen.

Jane's contribution is particularly telling in that it suggests that the purpose of the instructional activity for her was to calculate using known facts rather than to generate experientially real numerical relationships. More generally, most of the students' activity had lost its grounding in situation-specific imagery by this point in the discussion. Although those who responded were able to produce correct answers for the most part, none articulated quantitative relationships that were grounded in the original situation.

Ten days later, on November 18, Ms. Smith introduced the scenario involving the twins for a second time. The quality of the discourse differed markedly from that on November 8. Ms. Smith's reflection on that prior discussion led her to initiate changes in the nature of the discourse by (1) supporting the students' construction of imagery of the situation with figures and numerals drawn on the board, and (2) by asking students about quantitative relationships implicit in their calculations.

T: Think back to . . . think about the twins we talked about (draws two faces on the white board, side by side). Think about the twins that we have and imagine that . . . imagine that these twins have a set of markers. They each have a set of markers and their set of markers has ten markers in it (writes the numeral 10 under each face). Each one has a set of ten markers. Altogether, if you put both of their sets together (puts a plus sign between the two 10s), how many would you have?

Next, she asked:

T: But you know what? These twins are irresponsible. They can't keep up with their markers and you know what? They both lost 'em. They both lost a marker so now each one of them has only nine markers (writes $9 + 9$ under the two faces). So now how many do they have? Amy?

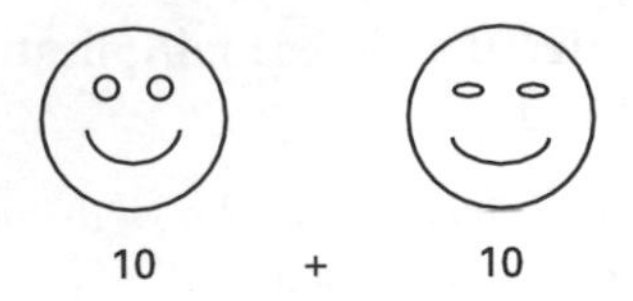

Figure 3.3. Twins shown with 10 markers each

Amy: I remembered that you had nine in one and nine in another and that would make eighteen.

Although Amy was not asked to explain how she knew $9 + 9 = 18$, her explanation appeared to be grounded in the situation of the twins as she talked about "nine in one [package of markers]."

Teri offered the next explanation:

T: Teri, did you think about it a different way?
Teri: If ten plus ten is twenty then nine plus nine can't be nineteen 'cause to have two nines you have to take two away.

Ms. Smith seemed to judge that other students might have difficulty in interpreting Teri's relatively sophisticated explanation in terms of relationships between quantities of markers. For these students, Teri's explanation would be a calculational procedure unconnected to the problem situation. Ms. Smith attempted to counteract this possibility by initiating a shift from the so-called mathematical register (Forman, in press) to discourse grounded in situation-specific imagery as she redescribed Teri's explanation:

T: Teri said they took away one marker here (points to one of the twins) and then took away another marker here (points to other twin) and that would be only eighteen left since two markers went away. They lost a total of two markers, they lost two whole markers so that means there would be eighteen. They lost markers number twenty and number nineteen so that left them with eighteen markers.

Although Teri's explanation was calculational, it could be argued that she had interiorized imagery (cf. Thompson, in press). She was one of only a few students who, in the initial interviews conducted at the beginning of the school year, used noncounting strategies flexibly to solve small number sentences. It therefore seems reasonable to suggest that she was able to reason numerically in a manner that maintained quantitative significance. However, this was not a possibility for many of the students.

As a consequence, for these students Teri's activity did not constitute an explanation of the problem as originally posed. Ms. Smith therefore initiated the *folding back* of discourse to make it possible for other students to interpret Teri's explanation in terms of situation-specific imagery. Although Teri's reasoning was relatively sophisticated, we note that the folding back of discourse would not constrain her activity in that she could interpret the discussion of the markers in terms of her outer-level knowing. At the same time, it served to support those students who were initially unable to interpret Teri's explanation in terms of quantitative relationships.

It is important to note that when Ms. Smith initiated the folding back of discourse she was simultaneously illustrating what, for her, counted as an appropriate explanation. In particular, in the explanation that she and Teri developed together, a numerical calculation expressed relationships between quantities in the task situation. Ms. Smith can therefore be seen to have initiated the renegotiation of the sociomathematical norm of what counts as an acceptable explanation in her classroom (cf. Yackel & Cobb, 1996). In doing so, she was inducting her students into a conceptual rather than a calculational orientation (cf. Thompson, Philipp, Thompson, & Boyd, 1994).

Dropping Back

As was the case in the first episode, numeral patterns also emerged as the focus of discussions in the second episode. In the first episode, Ms. Smith initiated the folding back of discourse to taken-as-shared imagery of the task situation. However, as such imagery had not been developed in the second episode, there was nothing to which Ms. Smith and the students could fold back. In this and similar situations, we speak of the discourse *dropping back* rather than folding back.

During October, a scenario involving a double-decker bus (van den Brink, 1989) was developed with the students with the intent that it would serve as a situation in which to investigate partitioning of numbers up to twenty (e.g., 15 partitioned as 15 and 0, 14 and 1, 13 and 2 . . .). The students were asked to think of different ways a certain number of people could be on the top and on the bottom decks of the bus. Ms. Smith first introduced the scenario of the double-decker bus on October 14, anticipating that the idea of a double-decker bus could become experientially real to the students as they participated in the narrative. However, initial discussions indicated that a double-decker bus was not real in the out-of-school experiences of most children in the classroom, as evidenced by the following exchange:

T: Now, we are gonna think about buses for a few minutes. Have you ever ridden on one of the Metro . . . one of the buses that will take you a long way down one street or down another street if your car is broken and you are needing to go someplace. Have you ever ridden a bus before?

Mike: I've ridden the [school] bus.

T: They're kinda like the [school] bus.

At this point several students share bus-riding experiences.

T: Well, some places have buses that instead of having just one set of seats like we have seats on our bus they have seats down on the bottom and then you can kinda go up stairs a little bit (shows picture of a double-decker bus on the overhead projector) to seats up on top. It's called a double-decker bus. Some towns have double-decker buses. And you can decide if you want to go up in the top or stay down in the bottom. Now think about a double-decker bus. Keep in mind when you get on you get to choose if you're gonna sit upstairs or downstairs. Some people might like to ride on the top and some people might like to ride on the bottom. Now, we're gonna think about a couple of things. When I ask you a question we're gonna pretend that the bus is empty . . . at the first stop and these are the first people getting on at the first stop where people . . . is there gonna be a problem where the people sit on the top or the bottom? Can they sit where they want? Dan?

Dan: I don't understand this double-decker bus. Is it like stairs like a house or do you have to climb up it? Does it have like a floor at the top?

Two students then offer descriptions of what the bus looks like. They have actually ridden a double-decker bus, one in England and one in Disney World.

T: Does anybody else have a question? Teri?

Teri: Is the top high up?

During this introduction to the double-decker bus scenario, Ms. Smith had attempted to support students' construction of situation-specific imagery by showing them a graphic of a double-decker bus and engaging them in stories about double-decker buses. However, it became apparent from Dan's and Teri's comments that the starting point of the double-decker bus activities was not experientially real for all of the students. This lack of taken-as-shared imagery gave rise to difficulties when Ms. Smith posed partitioning tasks.

T: Now, think about the bus and think about the first stop. The top deck has as many as need to get on or the bottom can hold as many as need to

Figure 3.4. Organizing chart for double-decker bus task

get on, but the people get to decide where they're gonna sit. So think about eight people are getting on at the first stop . . . if eight people are getting on at the first stop . . . how many people could . . . how could they sit on the bus . . . how could they sit on the top and the bottom . . . there are different ways the number of people could be on the top and different numbers could be on the bottom (draws an organizing table to record students' suggestions, see Figure 3.4). And if we're gonna think about eight people where could . . . now it doesn't matter which seats they are in but could they be on the top or could they be on the bottom? How many could be in each place? Amy?

Amy: Seven.

T: Where could seven be? You mean seven in each place?

Amy: Yes.

T: Now if we put seven in each place where are we gonna get all those people cause we only have eight people? So if seven get on the top Amy, how many get on the bottom. (She does not respond.) Kim?

Kim: One.

T: Why one Kim?

Kim: (Shrugs shoulders to indicate she does not know.)

It could be argued that at this point students had a range of differing and possibly incompatible interpretations of the task situation. Although the teacher and students were eventually able to generate all the partitionings of eight, the process became a purely calculational one of generating pairs of numbers that summed to eight. The students' approaches to the instructional task did not appear to grow out of the scenario and seemed to lack grounding in imagery. However, at the time, this was not apparent to either us as observers or to Ms. Smith. It is only with hindsight that we are able to question to what degree the students' activity was grounded in situation-specific imagery.

The next occasion on which the double-decker bus scenario was introduced occurred four days later on October 18. Initially, Ms. Smith attempted to help the students recall the prior discussions about the bus.

T: Remember the double-decker bus? A double-decker bus is a special kind of bus. Lynn, where can you sit on a double-decker bus?

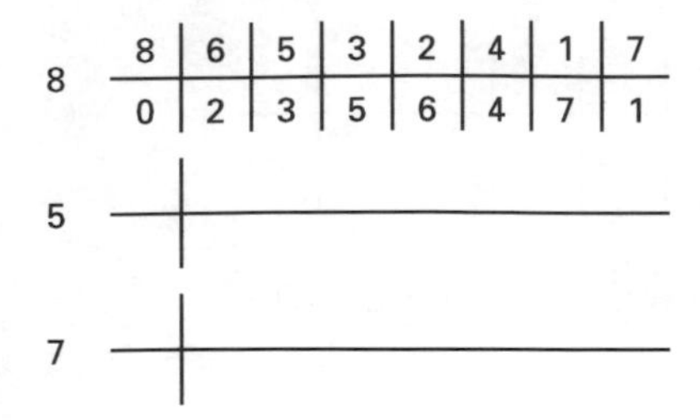

Figure 3.5. Sample activity sheet for double-decker bus activities.

Lynn: You can choose any of the seats as long as nobody is sitting in them.
T: You can choose any seat? Where could you go, you could go . . . something about the stairs?
Lynn: The top or the bottom.
T: Lynn says if you get on and if the bus is empty you can pick any seat in the top or any seat in the bottom.

Lynn's initial response in the initiation-response-evaluation sequence did not fit with what Ms. Smith intended and so she gave a suggestive hint by referring to "the stairs."

Next Ms. Smith introduced the task as follows:

T: Try to think of all the different ways . . . of all the ways . . . of all the ways they could decide to be on top or on bottom. Think of the different numbers that could be on top and the different numbers that are going to be on bottom.

Here, Ms. Smith appeared to pose the task in numerical terms, thereby supporting students' interpretation of this as a purely calculational task. The students then completed individual activity sheets where they recorded different partitionings of a given number of people on the bus. An example of one students' written work is shown in Figure 3.5. These solutions were then discussed in a subsequent whole class discussion.

Prior to the lesson, Ms. Smith commented that she hoped the students would begin to organize the partitionings recorded in their tables when they considered whether or not they had found *all* the ways. She explained that she intended to capitalize on solutions in which students recorded commutative partitionings or organized the table in some other way. She thus anticipated that students would be able to reflect on the results of their partitioning and discuss how they might organize the results. However, in her attempt to initiate this reflective shift in discourse, the discussions became calculationally oriented and lacked any explicit

7	0	6	1	5	2	4	3		
0	7	1	6	2	5	3	4		

Figure 3.6. Jake's solutions to 7 people on the bus

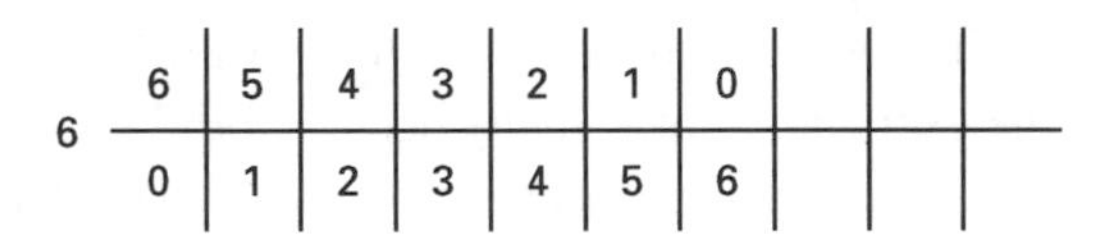

Figure 3.7. Dan's solutions to 6 people on the bus

reference to situation-specific imagery or to quantitative relationships (cf. Thompson et al., 1994). The calculational focus is apparent in the following exchange:

T: When I was walking around one thing I saw that people did . . . I saw was what Jake did. Okay, if Jake if you could explain and I'll write it down so if you could explain (Ms. Smith notates, see Figure 3.6).

Jake: I put seven on top and zero on bottom and zero on top and seven on bottom then six on top and one cause it still makes seven. Six on the bottom and one on top.

T: How did you decide to use those numbers? Why did you put seven first?

Jake: Because oh I guess I'll just put seven first and keep going backwards.

T: Okay, he put seven and did both of them then he went to six next and then what you have is five. Okay, tell us how what you mean that you started at seven and keep going backward.

Jake: Okay, seven, six, five, four (points to numbers) then three, two, one.

T: Jake thought about when he had both ways to use seven he went to the six and when he had both ways to use six he went to five. Is that what you did? Did that make it easier to find them all?

Jake: (Nods).

In this exchange, the discourse focused on patterns in the numerals that did not signify quantitative relationships in the bus scenario. Although the numeral pattern that emerged may have had quantitative significance for some of the students, the classroom discourse did not make reference to the people on the bus and may have been about numerals per se for many students. The calculational focus continued as students discussed Dan's solution for six people on the bus (see Figure 3.7).

Jon: 'Cause six, five, four, three, two, one, zero and six, five, four, three, two, one, zero. One goes that way and one goes that way (points to the two rows of numerals).

Mike: To make it easy you could just use a rocket countdown six, five, four, three, two, one, zero, blast-off!

Carl: There's a countdown on top and a countdown on bottom.

Ms. Smith presumably highlighted solutions which students had attempted with the intent of providing learning opportunities for other students. However, the exchange that actually occurred did not give rise to such opportunities. Numerical patterns rather than quantitative relationships emerged as the focus of discourse when Ms. Smith and the students discussed the "countdown" approach. When the students first completed the activity sheet, some might have related partitionings such as 6/0 and 5/1 by imagining a passenger moving from the top deck to the bottom deck of the bus. However, the whole class discussion appeared to mitigate against other students making interpretations that involved relating quantities in the situation.

After this lesson, the project staff and Ms. Smith met to discuss the instructional sequence and the double-decker bus problems in particular. It was at this time that we first realized that the double-decker bus scenario might not have been experientially real for the students. In other words, there might have been nothing to which either the students individually or the class collectively could fold back. This being the case, it would have been extremely difficult for Ms. Smith to guide the development of discourse that focused on anything but numeral patterns. The research team therefore discussed instructional activities in which the bus scenario might become experientially real for students. In addition, we also considered the nature of classroom discourse that would support students' construction of quantitative relationships. As a consequence of these discussions, Ms. Smith decided that, with the help of several students, she would act out situations in which passengers got on and off the double-decker bus. The following day, she explained to her students that they would pretend that a table top and the floor were the two decks on the bus. She then selected a group of three students and asked them to demonstrate the different ways that they could ride on the bus. Ms. Smith supported this activity by providing a commentary for the observing students.

After several possibilities had been enacted by different numbers of students, the students worked individually to complete double-decker bus activity sheets. As they did so, Ms. Smith intervened by asking them to explain their thinking.

T: Lynn, can you tell me what you were thinking about to decide how the people should be on the bus?

Instead of focusing on the numeral patterns they generated, Ms. Smith asked the students to explain their thinking in terms of quantitative relationships grounded in situation-specific imagery. Ms. Smith also supported the construction of such relationships as she guided shifts in discourse during the subsequent whole class discussion. In doing so, she recorded their solutions on the white board and attempted to ensure that the students' activity remained grounded in situation-specific imagery.

T: Okay, five people are getting on so five people get to decide if they are going to get on the top or on the bottom. Now let's think of ways they could make those decisions. How could they decide to sit on top and on bottom? Sue, what's one way they could decide?

Sue: Four people on the top and one people . . . one person on the bottom.

T: Okay, Sue said that four people decided to climb the stairs to the top and one decided to stay on the bottom.

Dropping back to the level of acting out the scenario appeared to support students' development of imagery and thus their construction of quantitative relationships when they both worked individually and explained and justified their solutions in the whole class discussions.

Dropping back can be contrasted with the folding back of discourse. In the case of folding back, collective activity at a more sophisticated level informs discourse at a revisited level. In a very real sense, it is not the same discourse even though, superficially, the same issues are being discussed (e.g., the number of markers that two twins have). In the double-decker bus episode, physically acting out the scenario cannot be characterized as folding back because it was not informed by prior discussions of partitioning numbers. The students sitting on the table and on the floor were passengers on the bus per se. The activity therefore constituted a new beginning and, in such instances, we speak of the discourse dropping back rather than folding back. The students did appear to construct imagery to which they could subsequently fold back as they participated in this activity. As a consequence, discussions focused on the creation of relationships between quantities in the scenario rather than on the generation of numeral patterns.

Taken-as-shared Basis for Communication

Thus far, we have differentiated between the processes of folding back and dropping back and have attempted to clarify the relationship between the two. In addition, we have stressed the importance of

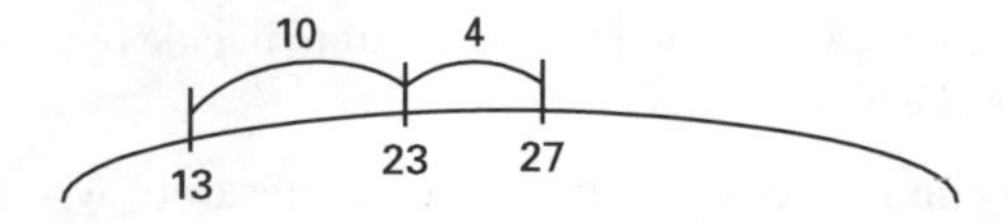

Figure 3.8. Record keeping on the empty number line

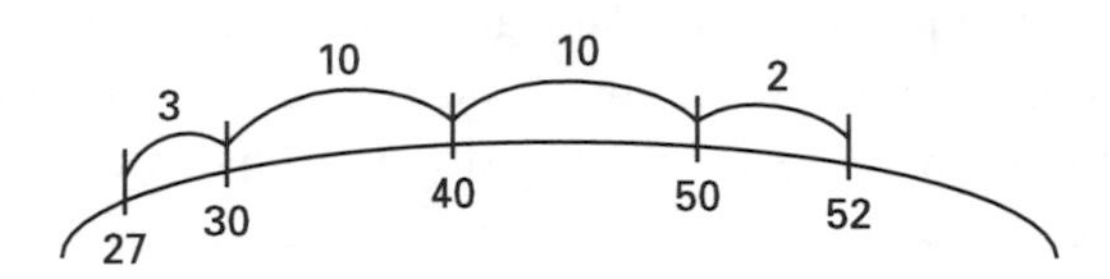

Figure 3.9. Twenty-seven candies in the shop and Ms. Wright makes 25 more

grounding students' increasingly sophisticated mathematical activity in situation-specific imagery. The third episode we will present also focuses on imagery. In this case, however, our concern is not solely with individual students' imagery, but it also encompasses imagery that is taken-as-shared by the classroom community. The episode illustrates the difficulties that arise when discourse folds back in the absence of taken-as-shared imagery.

During February, a scenario involving Ms. Wright's Candy Shop in which candies were packed in rolls of ten was developed with the students. They worked with unifix cubes as substitutes for candies and the initial instructional tasks included estimating, quantifying, and partitioning collections of candies. In March, the scenario was elaborated by introducing a number line "empty" of numerical increments as a device for recording transactions in Ms. Wright's Candy Shop (see Figure 3.8).

Initial activities included tasks in which students found how many candies Ms. Wright would have if she either made or sold a small number of candies. The empty number line was used to record the *results* of these transactions by incrementing or decrementing the appropriate number of candies. Later, Ms. Smith posed addition and subtraction tasks by drawing an empty number line and describing a transaction in the candy shop. Ms. Smith would then record the calculational *process* of students' solutions on the white board (see Figure 3.9).

As the instructional sequence progressed, we inferred that most of the students' activity continued to be grounded in imagery of the candy shop. One of our primary sources of evidence for this inference was the fact that students, working in pairs, were able to create appropriate empty number lines to tell stories that described a series of transactions in the candy

Figure 3.10. Ninety candies in the shop and Ms. Wright sells 8 of them

shop. However, a classroom episode that occurred on March 30 convinced us that, for at least some of the students, this had become a calculational task situated in purely calculational discourse.

In this lesson, Ms. Smith posed addition and subtraction tasks by drawing a horizontal empty number line and by describing transactions in the candy shop, but without acting them out. One of the tasks posed was *Ms. Wright has 90 pieces of candy and she sells 8 of them. How many candies does she have left?* (See Figure 3.10.) Teri and Dave justified their answers of 82 by giving collection-based and counting-based explanations, respectively. At this point, Bob justified his answer of 81.

Bob: I think it's eighty-one. Because if we're already down to ninety and then you don't count . . . you don't count ninety because if you have eighty and then you take away nine, it would get you down to eighty.

Bob argued that 90 should not be included in the numbers that you take away. For him, the numbers to be taken away were the eight numbers from 89 through 82 inclusively, leaving him with 81 since "we're already down to ninety."

As the episode continued, several other students offered a range of alternative numerical interpretations that conflicted with each other. A detailed analysis of the entire episode revealed that most of the students were reasoning quantitatively (Cobb, Gravemeijer, Yackel, McClain, & Whitenack, in press). However, the divergence in their individual interpretations and the problems in communication that ensued indicated that the imagery underlying activity with the empty number line was not taken-as-shared.[2] Further, simply referring to the candy shop scenario was insufficient and could not serve as a basis for imagery. Instead, the students had each individually given quantitative significance to their activity with the number line in terms of counting. It was not until this point in the sequence, however, that we appreciated the significance of counting as a source of imagery for the empty number line.

The range of interpretations that emerged is indicated by Dan's explanation for his answer of 83.

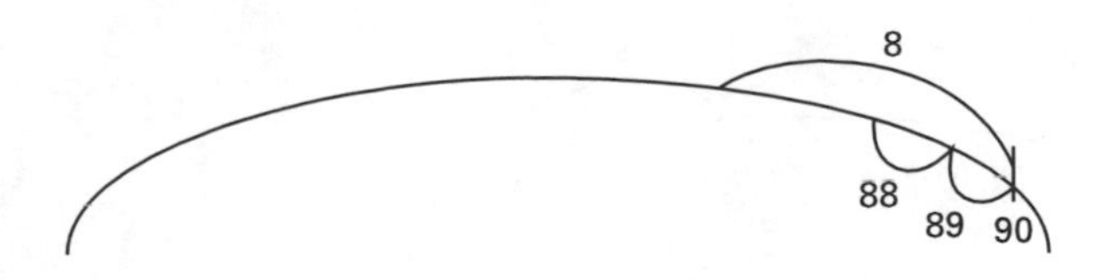

Figure 3.11. Symbolizing Dan's solution

Dan: I think it's eighty-three because I'm counting the ninety as a number.

T: Now we're talking about pieces of candy. There're ninety pieces of candy. Okay?

Dan: Okay, if you have ninety pieces of candy and . . . and it couldn't be eighty-two 'cause you'd be . . . 'cause the ninety, if you're taking away it and counting one, well it would just be the ninety-eight [*sic*]. It . . . you would be counting something extra. So you would take away two . . . the ninety-eight and the ninety-nine [*sic*].

Dan appeared to misspeak when he said 98 and 99 instead of 88 and 89. At first glance, it might seem that he had arrived at his answer of 83 by counting backward starting from 90 rather than 89. However, his comments in the remainder of the episode indicated that his interpretation of the task was relatively sophisticated. Collectively, these comments suggest that, for him, the ninth decade when counting comprised 80, 81, 82, . . . 89. The "ninety" to which he referred appeared to be of special significance in that it signified an additional item beyond this decade. By this reasoning, the solution to 90 – 8 involved taking away the 90 and then 7 from the decade 80, 81, 82, . . . 89. The result for Dan was then 83 rather than 82.

T: We're at ninety . . . we have ninety pieces. You said if you took away one of those pieces you would have . . . (notates, see Figure 3.11).

Dan: Eighty-nine. If you took away another one you would have eighty-eight. Now, you've got three pieces away. Now, you take away . . .

T: Now wait a second. You took away, Dan, you said you took away one piece, and that left you with eighty-nine. Is that what you said?

Dan: Right.

T: Then you took away one more piece, and that left you with eighty-eight. Is that right?

Dan: Right.

T: So how many pieces have you taken away so far?

Dan: Three.

T: Okay, show me where are the three pieces you took away.

Dan: The ninety, ninety-eight, I mean the ninety, the eighty-nine, and the eighty-eight (points to 90, 89, and 88 as he speaks).

For Dan, 3 rather than 2 candies had to be taken away to leave 88 because there was an additional candy beyond the ninth decade, "the ninety." Following this line of reasoning, 90 take away 2 would be 89 because it was necessary to take away the 90 and then one from the ninth decade. Although Dan's reasoning appeared to have quantitative significance for him, the lack of a taken-as-shared interpretation of the number line resulted in Dan and Ms. Smith talking past each other. Ms. Smith's recasting of questions in terms of candies was insufficient because there was a lack of shared interpretation of the number line itself, not of the original task involving candies.

At this point, Jake offered a solution that made it possible for numerical interpretations to become a topic of conversation.

Jake: Pretend this is ninety (holds up all 10 fingers).

T: Okay, pretend this is ninety (holds up her fingers). Ten, twenty, thirty, forty, fifty, sixty, seventy, eighty, ninety (flashes 10 fingers 9 times).

Jake: All right, if you take away one, do you see Dan (folds down 1 finger)? You take away one, two, three, four, five, six, seven, eight. Eighty-two. 'Cause see watch Dan. One, two, three, four, five, six, seven, eight. Eight (shows his fingers). Two more.

In holding up his fingers as countable objects, Jake folded back as he explained his reasoning. This individual act initiated a folding back of discourse in that other students referred to his fingers. However, this was not sufficient to establish a basis for communication even though the students fulfilled their obligation of attempting to understand each others' explanations. They made differing interpretations of a count of Jake's fingers just as they had of acting with the empty number line.

Bob: But that would leave the eighty and eighty-one 'cause I don't think y'all . . . I don't think y'all . . . Jake are you counting the eighty?

Jake: I'll show you, Bob.

Bob: You gotta count the eighty.

Jake: All right, this is ten (holds up 10 fingers). Pretend that's ninety. Are you agreeing with him (referring to Dan)?

Bob: I don't think it's eighty-three or eighty-two. I think it's eighty-one.

Jake: No. This would make ninety (holds up 10 fingers). One, two, three, four, five, six, seven, eight (closes 8 fingers as he counts).

Bob: But you've got to count the eighty, too.

Dan: But wait a minute. These are two (points to Jake's 2 fingers unfolded). Took away eight and got two, count the eighty, you have eighty-three.

As noted earlier, for Dan, the ninth decade comprised 80, 81, 82, . . . 89. Jake's two remaining fingers signified 81 and 82, but the 80 from this decade had not been taken away. Consequently, in his interpretation, there were 3 additional candies, those signified by 80, 81, and 82. Bob and Jake also seemed to assume that the ninth decade comprised 80, 81, 82, . . . 89. However, they appeared to reason that taking away 8 as indicated by Jake folding down 8 fingers would leave the candies signified by 80 and 81, so there would be 81 candies left.

The students' explanations appeared to carry the significance of acting on experientially real quantities. The difficulty arose however from the fact that Ms. Smith and the children interpreted the empty number line in a variety of different yet personally meaningful ways. They were unable to establish an adequate basis for communication during the remaining 10 minutes of the lesson. This was the case even though Ms. Smith redescribed several of the students' explanations in considerable detail by referring to the empty number line she had drawn. In the absence of taken-as-shared imagery for the empty number line, the situation remained irreconcilable.

In reflecting on the lesson, the project staff and Ms. Smith agreed that the students' explanations, while calculational, were grounded in the imagery of the situation. This judgment was based on the ways in which they attempted to justify their reasoning. We also noted the role that internalized or interiorized counting activity played in the students' explanations. In particular, the children had created a grounding for the empty number line individually and privately by interpreting it in terms of counting activity. In looking through the empty number line notation to counting activity, they had in effect imagined an activity that could have given rise to this way of symbolizing. Finally, we concluded from their difficulties in communicating that there was no taken-as-shared starting point to which the discourse could fold back.

As a consequence of this discussion, Ms. Smith decided that on the following day she would introduce unifix cubes as countable items in conjunction with the empty number line in an attempt to support the negotiation of taken-as-shared imagery. In doing so, she would be initiating the dropping back rather than the folding back of discourse. Previously, the empty number line had emerged as a way of recording transactions in the candy shop. Ms. Smith's intent was that it would now become grounded in taken-as-shared counting practices.

To this end, she arranged unifix cubes in bars of 10 of differing colors. She placed the bars end to end in a tray under the white board and drew an empty number line directly above them. This arrangement would

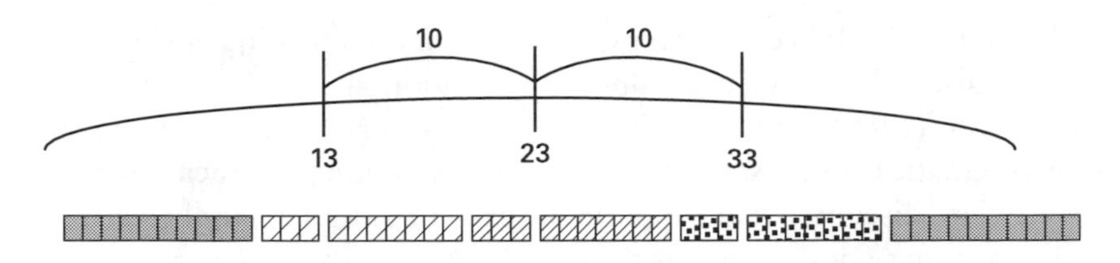

Figure 3.12. Supporting incrementing activity with cube train

make it possible for the students to refer to and, if necessary, count the cubes to explain their reasoning with the empty number line. Ms. Smith introduced the cubes as candies and cast all problems in terms of the candies. When students offered solutions to incrementing and decrementing tasks, Ms. Smith symbolized them on the empty number line and partitioned the train of unifix cubes at the appropriate points to correspond with the jumps on the empty number line (see Figure 3.12). Further, on occasion she specifically asked the children to determine which pieces of candy would be counted when going from, say, 34 to 40.

Several students modified their interpretation of the empty number line as a consequence of participating in these discussions. For example, Dan explained his solution to the task *Ms. Wright has 20 candies in the shop and she sells 2* by stating:

Dan: You take away the twenty and the nineteen leaves eighteen.

More generally, this dropping back of discourse made it possible for the students and Ms. Smith to develop a viable basis of communication. Although the students soon reasoned independently of the train of unifix cubes, they sometimes referred to it when justifying their thinking. In doing so, they initiated the folding back of discourse to unifix cubes that were, for them, no longer the same cubes as before.

Conclusion

Throughout this chapter, we have attempted to illustrate the role of imagery and discourse in supporting the development of mathematical understanding. We have distinguished between the folding back and the dropping back of discourse, and have discussed the importance of taken-as-shared imagery. The examples from Ms. Smith's classroom illustrate that students' activity can lose its grounding in imagery even when the explicit and consciously pursued goal is to support the growth of understanding. The sample episodes point to a tension that can sometimes arise between supporting reflective shifts in discourse such that what is said and done itself becomes an explicit topic of conversation, and ensuring that discussions remain grounded in situation-specific imagery. Pirie and Kieren (1989) argue that, in psychological terms, the development of mathematical understanding is a recursive, nonlinear phenomenon. In our view, discourse that supports the growth of understanding shares these characteristics. This was the case in Ms. Smith's classroom as she strove to ensure that her students' mathematical activity had quantitative significance.

Acknowledgment

The research reported in this chapter was supported by the National Science Foundation under grant No. RED–9353587. The opinions expressed do not necessarily reflect the views of the Foundation.

Notes

1. The teacher, Ms. Smith, was a highly motivated and very dedicated teacher in her fourth year in the classroom. She worked to create a learning environment that supported her students' learning and was attempting to reform her practice prior to our collaboration. Our collaboration with Ms. Smith resulted from mutual needs and interests. Ms. Smith was seeking help and guidance with her efforts at reform; we were seeking a teacher with whom to collaborate as we developed instructional sequences. In particular, the teacher viewed the other members of the research team as peers with whom she reflected daily as she attempted to reform her practice.
2. The initial instructional activities in the sequence, as it was originally outlined by its developers, involved the use of a bead string composed of 100 beads (Treffers, 1991). The beads were of two colors and were arranged in groups of 10. We decided to modify the sequence by omitting the instructional activities involving the bead string because we judged that the string did not serve as a means by which students might explicitly model their prior problem-solving activity. Instead, we decided to develop the empty number line sequence by building on the scenario of the candy shop. While unifix cubes had been used as models of the candies in the candy shop sequence, the cubes had not been used to establish a basis of imagery for the number line. In retrospect, we realized the importance of activities that support the establishment of a basis in imagery to underlie the number line and thereby make effective mathematics communication possible.

References

Cobb, P., Boufi, A., McClain, K., & Whitenack, J. (1997). Reflective discourse and collective reflection. *Journal for Research in Mathematics Education, 28* (3), 258–277.

Cobb, P., Gravemeijer, K., Yackel, E., McClain, K., & Whitenack, J. (in press). Mathematizing and symbolizing: The emergence of chains of signification in one first-grade classroom. In D. Kirshner & J. A. Whitson (Eds.), *Situated cognition theory: Social, semiotic, and neurological perspectives,* pp. 151–234. Hillsdale, NJ: Erlbaum.

Cobb, P., & Yackel, E. (in press). Constructivist, emergent, and sociocultural perspectives in the context of developmental research. *Educational Psychologist.*

Forman, E. (in press). Forms of participation in classroom practice: Implications for learning mathematics. In P. Nesher, L. Steffe, P. Cobb, G. Goldin, & B. Greer (Eds.), *Theories of mathematical learning.* Hillsdale, NJ: Erlbaum.

Gravemeijer, K. (1990). Context problems and realistic mathematics instruction. In K. Gravemeijer, M. van den Heuvel, & L. Streefland (Eds.), *Contexts, free productions, tests, and geometry in realistic mathematics education,* pp. 10–32. Utrecht, The Netherlands: OW & OC Research Group.

Gravemeijer, K. (in press). Mediating between concrete and abstract. In T. Nunes & P. Bryant (Eds.), *How do children learn mathematics?* Hillsdale, NJ: Erlbaum.

Lave, J. (1993). Word problems: A microcosm of theories of learning. In P. Light & G. Butterworth (Eds.), *Context and cognition: Ways of learning and knowing,* pp. 74–92. Hillsdale, NJ: Erlbaum.

National Council of Teachers of Mathematics. (1991). *Professional standards for teaching mathematics.* Reston, VA: National Council of Teachers of Mathematics.

Nunes, T., Schliemann, A. D., & Carraher, D. W. (1993). *Street mathematics and school mathematics.* New York: Cambridge University Press.

Pirie, S., & Kieren, T. (1989). A recursive theory of mathematical understanding. *For the learning of mathematics, 9* (3), pp. 7–11.

Saxe, G. B. (1991). *Culture and cognitive development: Studies in mathematical understanding.* Hillsdale, NJ: Erlbaum.

Streefland, L. (1991). *Fractions in realistic mathematics education: A paradigm of developmental research.* Dordrecht, The Netherlands: Kluwer.

Thompson, A. G., Philipp, R., Thompson, P. W., & Boyd, B. (1994). Calculational and conceptual orientations in teaching mathematics. In *1994 Yearbook of the National Council of Teachers of Mathematics,* pp. 79–92. Reston, VA: National Council of Teachers of Mathematics.

Thompson, P. W. (1992). Notations, principles, and constraints: Contributions to the effective use of concrete manipulatives in elementary mathematics. *Journal for Research in Mathematics Education, 23,* 123–147.

Thompson, P. W. (in press). Imagery and the development of mathematical reasoning. In L. P. Steffe, B. Greer, P. Nesher, & G. Goldin (Eds.), *Theories of learning mathematics.* Hillsdale, NJ: Erlbaum.

Treffers, A. (1991). Didactical background of a mathematics program for primary education. In L. Streefland (Ed.), *Realistic mathematics education in primary school,* pp. 21–57. Utrecht, The Netherlands: CD-ß Press.

van den Brink, F. J. (1989). *Realistisch rekenonderwijs aan jonge kinderer.* Utrecht, The Netherlands: Vakgroep Onderzoek Wiskundeonderwijs & Onder Wijscomputer Centrum, Rijksuniversiteit, Utrecht.

Walkerdine, V. (1988). *The mastery of reason.* London: Routledge.

Yackel, E., & Cobb, P. (1996). Sociomathematical norms, argumentation, and autonomy in mathematics. *Journal for Research in Mathematics Education, 27,* 458–477.

4 Building a Context for Mathematical Discussion

Rodney E. McNair

A question that the mathematics education community is currently addressing is how to design mathematics classroom instruction that supports and encourages students' active, authentic engagement in the construction of mathematical knowledge. Creating authentic problems is one strategy that is often considered in this respect; however, authentic mathematical activity involves more than just the kind of problems students may be given to solve. We also need to consider the cultural authenticity of classroom instruction. Authentic mathematical activity involves students' adopting perspectives, beliefs, values, and expectations consistent with those of the mathematics community, and using these to analyze a problem situation with respect to their existing mathematical knowledge and experiences.

A major part of doing mathematics involves interpreting situations according to the cultural norms established by the mathematics community. To the extent that students are able to use these norms to analyze situations and to create the problem-solving contexts that they work in, their participation in mathematics classroom knowledge-construction activities can be said to be more or less authentic. While mathematicians may work in a context first established by other mathematicians, possessing the skills necessary to create a mathematical context is a necessary requirement for interpreting the quality and validity of such a prefabricated context. Traditional mathematics instruction typically provides the goal and the numbers and simply requires the student to execute a procedure selected from a restricted set of options (Cognitive and Technology Group at Vanderbilt, 1993). Traditional mathematics instruction also typically provides the words and the syntax within a restricted context that limits students' discourse choices. Both of these situations must be addressed by new mathematics curricula and instruction.

In this chapter, I will discuss the discourse requirements and constraints imposed by different kinds of mathematics problems and the possible effects they may have on students' ability and inclination to create their own mathematical contexts. I will consider three different mathematics classrooms with respect to the kinds of problems students are asked to solve and the role of the teachers and the students in the construction of a context that supports the development of mathematical knowledge and discourse. The purpose here is not to compare the classrooms or the instruction in the examples to suggest that one situation is somehow better than another. The goal here is to add to our understanding of how we might design instruction to transition between the distinct situations represented in the examples, in a manner that supports and encourages the development of students' mathematics discourse skills.

In each example, I will describe the relationship between the contexts presented by the teachers and the contexts developed by the students with respect to the opportunities that are provided for the students to develop cultural views and a discourse that are consistent with those of the mathematics community. These examples are also used to describe the discourse competencies students will need to develop through classroom instruction in order to be authentic participants in authentic mathematical activities.

Texts As Knowledge Constructing Contexts

An analysis of the kinds of discourse practices and the resultant texts that students and teachers develop in mathematics classrooms can inform our understanding of mathematics teaching and learning processes. The term *text* is used here to refer to the verbal portion of a discussion that can be recorded and analyzed within the context of the situation in which it occurred. A text is a series of utterances tied together to form a cohesive system that allows for relatively consistent interpretation (Brown & Yule, 1983). During problem-solving activities, it is the text of the discussion that students can reflect on to support their thinking. The term is used here instead of terms like *discourse* or *discussion. Discourse,* as used in this chapter, refers to the sociocultural and cognitive processes that produce an utterance, while discussion is used to describe what it is that people are doing when they talk to each other.

Texts form a part of the remembered experience we take from a problem-solving situation. As present and past texts are linked they may form a more or less coherent context to support problem solving. Within this framework mathematics can be thought of as a grand text, shared, developed, and maintained in classrooms and in the mathematics community. Each mathematical experience adds to the individual's text, and as these experiences are shared and compared with the texts of others, new texts are formed and existing texts are passed on.

One of the goals of the mathematical communication process is to create an abstract context in which mathematical arguments can be constructed and resolved within a coherent system of assumptions and theories. It is in these abstract contexts that mathematical arguments can be carried to their full conclusions and result in the development of meaningful mathematical knowledge. The resulting knowledge can then be applied to real-world problems by recontextualizing the mathematical results. Learning to communicate mathematically and learning mathematics through communication depend on the students' and the teachers' ability to create and work within these mathematical contexts.

In the social process of abstracting problem situations and experiences, texts are created consisting of the participants' utterances. Such texts form a major part of the context in which mathematical discussions take place, and constructing such texts plays a major role in the doing and learning of mathematics. These texts may refer to the texts of previous discussions and may be referred to in future discussions. The kind of texts that are created as a result of mathematics classroom instruction will depend on the problems teachers pose for their students and the roles that teachers and students play in the solution process.

Mathematics Frame

One of the results of students' long-term participation in mathematics classroom activities is the development of a mathematics frame. Frames are developed through experience, and so the kind of experiences students have in the mathematics classroom will determine the kind of mathematics frame they will develop. As students engage in mathematics classroom activities they must decide what they are doing, why they are doing it, and what they should expect of themselves and others. It is their mathematics frame that will guide the decisions they make, and that will guide their text construction process.

The term *frame* is used here to refer to the set of epistemological, ontological, and emotional characteristics we use to interact with each other and our environment (McNair, 1996). Frames are socially constructed expectation structures that we use to structure a new situation in terms of past experiences and to produce and interpret any communication associated with the situation (Goffman, 1975). Shared frames are developed as a group of people share a common set of experiences in some restricted domain over an extended period of time. As these groups grow in size, they may form communities that develop their own cultures, complete with their own beliefs, values, language, discourse, history, and institutions.

Over the years mathematicians have come together to form a community that shares a common interest in the domain of mathematics, and as a result they have developed their own culture and their own frame. The beliefs, values, and expectations that underlie the *mathematics frame* include the objectivity of number, the logical ordering of the universe, and the objective role of number in that ordering. Value is placed on precision, definition, dichotomy, logic, and abstract objectivity. Expectations would include the giving and receiving of explanations and proofs, and an emotional detachment from these explanations and proofs (McNair, 1996). Learning to do mathematics requires becoming a member of the mathematics community, at least to the extent that one can think like and talk like a mathematician, which in turn requires that cognitive and communicative processes be driven by a mathematical frame. The joint process of thinking and talking is referred to here as discourse, and when discourse is driven by a mathematical frame the result is referred to as mathematical discourse.

Other Frames. Discussions are not carried out in one frame, but consist of many frames interacting within the ebb and flow of the conversation (Goffman, 1975; Pirie, 1991). Additional frames that may play significant roles in mathematics classroom discussions include the *problem frame,* the *calculation frame,* and the *peer frame* (McNair, 1994). The *problem frame* is the intuitive, concrete frame of reference provided by the statement of the problem. Utterances in the problem frame take their referentially semantic content directly from the concrete relationships described by the problem. In the problem frame constraints and goals defined by the real world context of the problem determine the appropriateness and validity of an utterance. For example, in a business problem the rules of supply and demand or efforts to increase a profit margin are what drive problem-solving activities. These are concepts that can be modeled using mathematical symbols and operations, but they are not mathematical concepts. In a business problem, it is business laws that must be understood and satisfied and mathematics is a tool that can assist in this effort.

In the *calculation frame* the referentially semantic content is the steps in a process, or the results of those steps, used to calculate an answer.

These processes are well-developed procedures that may have required mathematical reasoning at some point in the user's education, that have become well-practiced routines. For example, for high school students arithmetic operations would be considered to be carried out in the calculation frame. These are short-term intermediate processes that provide additional information to be used in either the problem or the mathematics frame.

Utterances that repeat or deal directly with the referentially semantic content of a peer's utterance are based in a *peer frame.* These utterances often request an explanation, request clarification, or build on what was said by the producer's peers. In the peer frame, the common subject and purpose of the group is the context in which the validity of statements is judged. As students begin to reflect on the utterances of their peers, they should begin to use these to frame their own utterances. This would lead to a group frame which would represent the shared understandings of the group. An increase in utterances based in the peer frame would also indicate a more reflective group discussion.

Frame and Text Development. Since texts are developed according to the frame used to interpret and guide the social interaction process (Goffman, 1975; Tannen, 1993), the kind of mathematical frame students develop will determine the kind of texts they will produce. The frame used to create and interpret the text will restrict the subjects and the purposes of the texts to those most commonly associated with that frame. A mathematics frame will restrict the discourse to a particular set of experiences which form the referentially semantic content of the frame (McNair, 1996).

A major part of problem solving involves the construction of a text which constitutes a cognitive context in which some solutions rather than others can be constructed, compared, and validated, and in which some tools are more appropriate than others. What I would like to argue here is that just as knowing mathematics is doing mathematics, developing mathematical frames involves, and requires, the construction of mathematical texts. To develop the discourse of mathematics, students must engage in the same kinds of activities that mathematicians do, including the application of appropriate frames to the abstraction of real-world contexts, and the analysis of mathematical contexts, that lead to the development of coherent mathematical text, models, and systems. The extent to which students participate in the text construction process and the way this process is scaffolded by the teacher may determine the level of authenticity associated with classroom activities and has implications for students' learning and discourse development.

Problem Types

For the purposes of this discussion, problems are placed in one of two categories. Type 1 problems are abstract mathematical questions where the context is restricted to numbers, shapes, and spaces and the relationships between them. Type 1 problems have a restricted context that constrains and defines the frame that students can use to develop their texts. Type 2 problems are contextual problems where the mathematical problem and the mathematical context must be abstracted from a real-world problem context. Type 2 problems can support the development of many different mathematical and nonmathematical texts since there are fewer constraints placed on the frames that students and teachers might use.

Type 1 and Type 2 problems provide different kinds of learning opportunities and place different demands on the students' ability to create mathematical texts. Due to their abstract nature, Type 1 problems restrict the choice of frames and provide the mathematical context and much of the mathematical text necessary to support mathematical discussion. For example, Pirie and Kieren (1992) discuss an activity in which students constructed knowledge of fractions by folding paper. In such activities, the texts of classroom discussions are restricted to the relationships between the fractional sections of the folded paper which establish a concrete referent for the mathematical symbols that represent the part whole relationship. Lampert (1988) provided her students with a similarly mathematical context through the investigation of bases and exponents. Both examples will be discussed in more detail later in the chapter.

Unlike Type 1 problems, Type 2 problems involve a minimum of two contexts. First, there is the real-world context which may involve business, science, art, or history questions, that have no particular relevance to mathematics or the construction of mathematical knowledge. These contexts are important because they allow students the opportunity to connect mathematical knowledge to real-world referents, potentially establishing a fuller meaning for mathematics concepts, by providing important connections between mathematics and the students' lives.

The second context associated with Type 2 problems, which should be the real focus of mathematics classroom activities, is the mathematical context. Due to the nature of Type 2 problems, the mathematical context associated with the problem must be abstracted and constructed by the student. The mathematical formulation of a Type 2 problem is constructed when the perceivable mathematical relationships between key elements of the real-world situation are recontextualized within the past mathematical experiences of the problem solver, thereby creating a context for forming mathematical texts. In this process, the text which serves as a context for meaning making is shifted from one that refers to real world objects and goals to one that refers to mathematical ideas and concepts. This shift in texts also requires a shift in frame.

There may be more than one mathematical context embodied in the real situation of a problem, as well as multiple real world contexts; however, there is always at least one of each, and so there are always at least two levels of text, which may support two different activities, that lead to the construction of two different kinds of knowledge. An important issue to consider is that finding answers to Type 2 problems does not always require the development of a mathematical text. This may occur since mathematical symbols and concepts can be a part of nonmathematical texts which lead to the solution of business, engineering, or chemical problems. The mere presence of mathematical symbols and words does not make a text mathematical.

For example, computing a fair interest rate for a loan does not require a mathematical text. It only requires the use of some business formulas and some well-practiced calculation skills. The formula $I = RT$ is not a mathematical concept. It is a mathematical model of a business concept. We can use the formula to study the properties of equations by generating the equivalent expressions $R = I/T$ and $T = I/R$, but the formulas themselves are business concepts that have meaning in a business frame. From a mathematical perspective $I = RT$ and $D = RT$ are the same and obey the same laws where there are no restrictions on the size of R, but in business and in physics R is governed by different laws. A discussion of the economic or physical limits imposed on R would not be a mathematical discussion, but such a discussion might use lots of mathematical symbols and words.

Type 1 and Type 2 problems may have very different implications for student learning. In Type 1 problems the mathematical context is provided and so may not become an issue of concern or discussion. If the teacher provides an abstract context for the students, then the context is based on the teacher's past experiences and the teacher's frame, which may or may not be shared by the students. If the students abstract the problem and build their own problem-solving context, then the knowledge construction process is based on the students' past experiences and on their frames. Text can be created in either case; however, the creation process may differ significantly depending on the problem type and the roles played by the teacher and the students.

It is unlikely that instruction would consist exclusively of Type 1 or Type 2 problems, and it is not my intention to suggest that one should be emphasized over the other. A mixture of both types is likely to make the most sense. The argument here is that as we consider the characteristics of the problems we assign in mathematics classrooms with respect to the content we would like to teach, we also need to consider how those problems may or may not support the development of mathematics discourse. As we choose problems to challenge students' conceptual understanding, we also need to choose problems that challenge their discourse competency. A key consideration for instructional design is that Type 1 and Type 2 problems may provide different opportunities for students to develop mathematics discourse and the inclination to use it during problem-solving situations.

If students are to develop a mathematical discourse they must be authentic participants in authentic discussions about authentic mathematical problem-solving activities (Schoenfeld, 1992). Authentic participation in a mathematical discussion requires proper subject, purpose, and frame alignment (McNair, 1994). If students do not have mathematical subjects, purposes, and frames, then they are not engaged in mathematical activity and they will not develop and use mathematical discourse; however, if students do adopt mathematical subjects, purposes, and frames, then the development of mathematics discourse should follow the same natural processes that led to its development by mathematicians.

In many cases when teachers supply an abstract problem context, the students' reasons for engaging in the activity may be based on meeting the needs and requirements set by the teacher instead of more intrinsic reasons the students might develop. In this case, the teacher attempts to provide the subject, purpose, and the frame for classroom activities and discussions. Like knowledge, however, subject, purpose, and frame cannot be transmitted in whole from the teacher to the students, and so these activities may provide fewer opportunities for students to develop mathematical frames. Allowing students to construct their own context may provide more opportunity for the development of mathematical frames and discourse. If students are truly going to engage in authentic mathematical activity, then they will need to be able to develop appropriate mathematical contexts that support the development of mathematical texts where mathematical concepts can be developed.

Establishing a Mathematical Context: Data Examples

I will present and discuss data from three classrooms to attempt to clarify the teachers' and the students' roles in the development of mathematical texts that support the construction of mathematical knowledge. These examples are discussed with respect to the kind of scaffolding and the kinds of discourse requirements imposed and the opportunities that are provided by the problem contexts defined by the teachers. The first two examples will show how the teachers' efforts to provide mathematical contexts that allow for active student investigation of mathematical concepts and processes can support the development of mathematical texts. These examples are based on Type 1 problems that place restrictions on the students' discourse that make it more mathematical; however, the kind of restrictions imposed in these examples and the results are very different. The third example is based on a Type 2 problem.

In the first example, the students worked within the Type 1 problem context provided by the teacher to establish a text for solving and discussing their problem. Here the context provided by the teacher scaffolded the students' text construction process by restricting the discussion to abstract mathematical concepts. In the second example, students were given a Type 2 problem that was closely related to the Type 1 context on which the students' prior instruction had been based. Here the similarity between the abstract characteristics of the problem context and prior instruction provided by the teacher provided a different kind of scaffolding for the students' text construction efforts.

The third example will be a sample of student discourse in a Type 2 problem context where the teacher did not provide explicit mathematical restrictions on the students' discourse. In this case, the scaffolding took the form of whole class discussions of the students' problem-solving experiences where the teacher was careful not to introduce any formal mathematical concepts or procedures. These discussions were intended to help the students organize their experiences to support their problem-solving efforts.

Creating Joint Contexts

Lampert (1988) presents an example of a Type 1 problem environment in which the teacher gives the students more responsibility for the mathematical context on which classroom instruction is based. This added responsibility added to the kinds of discourse opportunities provided by the Type 1 problem. Discourse opportunities were also increased by the fact that the students were answering open-ended questions instead of simply searching for a calculation to use to find an answer. As Lampert argues, the added responsibility and the open-ended questions helped to make the problem-solving activity more consistent with real authentic mathematical behavior.

In this first example, the teacher provided a context which the students modified, and then the teacher adopted the students' modified version as her own to continue the lesson. "During the activity they invented a way of thinking about relationships among the numbers upon which I could build to take them into new mathematical territory" (Lampert, 1988, p. 19). Lampert's students had been asked to discover patterns in the numbers from 1 to 100 raised to the second power as part of a classroom activity. Her students found and proved the existence of patterns in the last digits of these numbers, and although the last digit may be the wrong digit to be concerned with in terms of exponential growth, the students' shift in context was able to support the development of some basic properties of exponents.

The students had just explained why the last digit in 7^4 was 1, by arguing that the result was obtained by multiplying 7^2 by 7^2, and they had been asked what they thought the last digit of 7^5 would be. Several students answered immediately:[1]

Arthur: I think it's going to be 1 again.
Sarah: I think it's 9.
Soo Wo: I think it's going to be 7.
Sam: It is a 7.

The teacher wrote these responses on the board and prompted the students to give justifications for their conjectures. This allowed the students to base their reasoning on their own experiences within the context that they and the teacher had created. Within this restricted context the students were able to create a text that supported their problem-solving efforts. They built on previously constructed texts concerning exponents, multiplication, and the patterns they had found in the last digits of the numbers. The following discussion took place as the students attempted to justify their conjectures.

Teacher: Arthur, why do you think it's 1?
Arthur: Because 7^4 ends in 1 and then it's times 1 again.

Gar: The answer to 7^4 is 2401. You multiply that by 7 to get the answer so it is 7×1.
Teacher: Why 9 Sarah?
Theresa: I think Sarah thought the number should be 49.
Gar: Maybe they think it goes 9, 1, 9, 1, 9, 1.
Molly: I know it's 7 'cause 7 . . .
Abdul: Because 7^4 ends in 1, so if you times it by 7, it'll end in 7.
Martha: I think it's 7. No I think it's 8.
Sam: I don't think it's 8 because it's odd number times odd number and that's always an odd number.
Carl: It's seven because it's like saying $49 \times 49 \times 7$.
Arthur: I still think it's 1 because you do 7×7 to get 49 and then for 7^4 you do 49×49 and for 7^5, I think you'll do 7^4 times itself and that will end in one.
Teacher: What's 49^2?
Soo Wo: 2401.
Teacher: Arthur's theory is that 7^5 should be 2401×2401 and since there is a 1 here and a 1 here . . .
Soo Wo: It's 2401×7.
Gar: I have a proof that it won't be a 9. It can't be 9, 1, 9, 1, because 7^3 ends in a 3.
Martha: I think it goes 1, 7, 9, 1, 7, 9, 1, 7, 9.
Teacher: What about 7^3 ending in 3? The last number ends in . . . 9×7 is 63.
Martha: On . . .
Carl: Abdul's thing isn't wrong, 'cause it works. He said times the last digit by 7 and the last digit is 9, so the last one will be 3. It's 1, 7, 9, 3, 1, 7, 9, 3.
Arthur: I want to revise my thinking. It would be $7 \times 7 \times 7 \times 7 \times 7$. I was thinking it would be $7 \times 7 \times 7 \times 7 \times 7 \times 7 \times 7 \times 7$.

Lampert's students were successful in finding patterns and in doing so they made implicit assumptions about the properties of exponents and also discussed some explicit procedures for calculating with exponents. In this case the students were able to work within the abstract context provided by the teacher to create their own abstract text, based in their knowledge of multiplication, to discuss number patterns and verify exponential operations. However, if we ask what made their discussion mathematical, the answer is largely the original context provided by the teacher which provided the subject, the purpose, and the language, and restricted the frame the students could use.

In this example, the teacher and the students created a text supported and restricted by the context originally established by the teacher. The students continued to work with exponents as instructed by the teacher;

however, their focus on the last digit was the result of insights they developed within the classroom community. Within this extended context the students were able to create a mathematical text, with the aid of the teacher, in which their subject and purpose were both restricted by the mathematical context established by the abstract concept of exponents and multiplication. All of their utterances must be interpreted and validated within the context of the rules for exponents and the multiplication process, and so they are restricted to a mathematical frame. The result is that the students' discourse and their discussion are both mathematical, but not necessarily because of any conscious efforts or intention on their part to make them mathematical. The abstract context provided by the teacher shaped the text by restricting the discourse choices available to the students.

While the scaffolding measures used by the teacher in this example may be appropriate, we must question the impact such scaffolding may have on students' development of the ability to create a mathematical text under less supportive conditions. In this example the context provided by the teacher shaped the students' discourse, but in order to do mathematics students must develop discourse skills that allow them to shape the context. Students must be able to create an abstract world of realities together with a set of assumptions, a set of declarative statements about the relationships that exist in their abstract reality, and a method to prove the validity of these statements (Snapper, 1979). Lampert's students did these things but their efforts were essentially extensions of the abstract reality provided by the teacher's initial question about exponents. The teacher's scaffolding does not minimize the mathematical quality of the students' problem-solving activities. In fact it may enhance it in many cases. However, we must consider how we might move beyond such scaffolding to provide greater discourse challenges and opportunities for students.

The Teacher's Context

Pirie and Kieren (1992) discuss data collected in classrooms with 8-year-olds and 12-year-olds working on the topic of fractions in order to demonstrate the learning characteristics and diversity of possible learning outcomes of a constructivist learning environment. In the episodes that they reported, the teachers built a mathematical context for discussing fractions that supported the students' active investigation of their understanding of fractions of the form $m/2^n$. The context built by the teachers consisted of paper repetitively folded in equal halves by the students, a kit containing unit, half, fourth, eighth, and sixteenth pieces, and activities designed to have the students search for patterns using the manipulatives as referents.

The restricted mathematical context established by the isomorphic relationship between the manipulatives selected by the teachers and the abstract concept of fractions facilitated students' learning. Students successfully worked in this context to investigate the properties of fractions and to develop understanding of equivalent fractions and equivalent combinations of fractions. The students and the teachers were able to create several mathematical texts based on the abstract problem context established by the teacher.

The following dialogue demonstrates how the students' work in this abstract Type 1 environment supported their efforts to develop a mathematical problem-solving text while solving a Type 2 problem. In this episode the students were given a work sheet containing the following problem:

Put eggs in a carton so that each of four persons gets a fair share. Color in the diagram with four colors showing each person's share.

The important thing to note here is that the problem is presented in a real-world context, different from the mathematical context in which the students had been working, and that the problem can be solved without knowledge of fractions. It is also important to note that there is a lot of similarity between the abstract characteristics of the problem and those of the instruction the students had received prior to this episode. As we will see, one of the students was able to recontextualize the problem in the context that they had been working in with the teacher, even though it was not required to find a solution to the problem they were solving.

Cathy: So how many do you have to color for each person?
June: How many each? (Looking and counting after sharing.) Three.
Sheila: I know each one gets one fourth!
Cathy: So how many eggs are there in one fourth?
Sheila: See (pointing to her diagram). Three. One, two, three. One, two, three . . .

At this point Sheila and Cathy were working in two different contexts. June had already answered the question by simply counting to find an answer, and so the relevance of Sheila's comment at that point was not immediately clear. What followed was a demonstration of how a Type 2 problem can lead to the establishment of a mathematical context and a mathematical text that has little to do with the original problem.

June: Three is what fraction of a dozen?
Sheila: I think I explained this a little bit harder. I took out one half and split halves into one quarter of a dozen.
June: You could just put in half then, 'cause it's half (relating result to halving action).
Sheila: But you already had half before you split it again.
June: When you split half into half what do you get?
Sheila: Only one fourth. I split the halves into one fourths.
June: But would that still be right here (with egg cartons and not paper or pizzas)?

The first two lines of dialogue above seem to deal with the real-world problem of counting eggs and are not produced by a mathematical discourse. However, in the third line the discourse and the text became mathematical and the students began to make utterances and to use language for mathematical reasons. At this point the text of the discussion formed a context in which the students investigated the meaning of half. The students' discussion might have continued to lead to the formation of an abstract definition of the concept of half, to eliminate the confusion caused by Sheila's ambiguous reference to the term.

The last utterance by June demonstrates that she had switched contexts from the egg carton problem to the abstract context the teachers had created and back again. The girls were having a mathematical discussion facilitated by the mathematical context that had been established in the classroom by the manipulatives and their prior problem-solving activities. The girls' discussion about halving halves has a mathematical subject and purpose. The subject has nothing to do with eggs, but, instead, it is the mathematical result of taking half of a half, the result of which hinges on whether the half is half of a whole, or is itself viewed as a whole. These are abstract halves and wholes, independent of the egg carton. The purpose of this brief discussion was to establish a shared mathematical understanding of a generalized halving procedure. All of the utterances are validated by an appeal to mathematical relationships and are intended to be interpreted in a mathematical frame.

The girls were able to create and work with an abstract representation of the egg carton problem to create a mathematical context that supported their development of a mathematical text. Their efforts were scaffolded by the similarity between the abstract relationships in the egg carton problem and the mathematical context that had been created by the teacher during prior instruction. Again, this scaffolding does not minimize the mathematical quality of the students' discussion; however, while selecting a Type 2 problem that matched prior instruction so closely was appropriate to this teacher's goals in this situation, making such a selection may not always be possible or desirable.

The scaffolding provided by the teacher in this case, as in the last example, is appropriate, but if students are to become autonomous users of mathematics and develop mathematical discourse skills, they must take more responsibility for the creation of the contexts that support their discussions, and teachers need to provide more opportunities for students to develop and practice their mathematics discourse skills. Unlike the last example where the students' discourse was restricted by the context provided by the teacher, in this example it was the similarity between the abstract context provided by the teacher and the egg carton problem that led the students to choose to create and to work in a mathematical context. The end result was still that the abstract context provided by the teacher shaped their discourse by limiting the discourse choices that were available.

This shift was not necessary or possible in the last example due to the abstract nature of the problem context. It is probable that those students could have made a similar shift if a Type 2 problem had been constructed to take advantage of texts the students had already developed concerning exponents. However, we must consider how we might move beyond this kind of scaffolding as well, so that students can create mathematical texts in situations that have not been specifically designed and chosen by the teacher. Also, it is difficult to imagine a Type 2 problem with the same kind of isomorphic relationship to the context involving exponents that Lampert's students developed in the previous example. As concepts become more complex and abstract, it may become more difficult to find such isomorphic Type 2 problems, so students will have to learn how to impose their own discourse restrictions on their text-building processes.

Student Created Contexts

Lampert's students extended the teacher's context to form a different abstract context based on the original, while the students in the Pirie and Kieren study were able to recontextualize a real world context into a familiar isomorphic mathematical context established by the teacher. In both cases the restricted context created by the teacher provided a mathematical restriction on the students' subject, purpose, and frame, which made their discourse mathematical and guided the text construction process. In this last example, I will examine the efforts of students to create a mathematical text in the absence of a teacher generated model to place restrictions on their discourse.

The data in the next example was taken from a study conducted by the author (McNair, 1994) in which the researcher served as the teacher during a 4-week summer intervention program for 18 students entering the ninth grade. Pretest results indicated that the students had not been formally exposed to the concepts to be discussed during the project. The goal of this project was to study students' mathematical discourse, and so there was a conscious effort not to impose any mathematical restrictions on the students' discourse. While all of the problems the students worked on during the project could be modeled by the equation of a line, there was no discussion of slope, intercepts, or independent or dependent variables that would have provided a restricted context for the students' problem-solving and text construction activities.

The goal of the instructional sequence used in the project was to get the students to develop the concepts associated with the formal equation of a line through group discussions with their peers and whole class discussions led by the teacher. As expected, all of the concepts necessary to establish this formal mathematical context were implicit in the students' group discussions, and the teacher's goal during the whole class discussions was to help the students make these concepts explicit through reflective discussion of their group problem-solving activities.

The teacher's subject during whole class discussions was the students' thinking and the purpose was to develop formal problem-solving rules and procedures based on the students' experiences. In this way the teacher provided a mathematical subject, purpose, and frame for the whole class discussions but did not provide a formal mathematical context. The context of the discussions was what the students had done and not the teacher's abstract formal mathematical representation of what they did or should have done. In this way the goal was similar to Lampert's efforts to build on the students' thinking; however, unlike the Lampert study, there was no a priori mathematical context which the students could extend to create their own version. The goal here was for the students to do mathematics and for the teacher to guide, but not to map, their journey. The teacher's role was to assist the students by asking questions that encouraged the students to reflect on the mathematics that was implicit in their discussions. This would require the students to make the same shift that Sheila made in the Pirie and Kieren study, but without the aid of a highly structured isomorphic mathematical context to assist them.

Although they had been told that the purpose of the summer course was to develop a formal problem-solving procedure, the students did not adopt this as their purpose. The students' discourse was consistently driven by a focus on finding an answer within the context of the real-world problem situations they were given. As a result of their focus on answers, their cooperative group discussions were based on nonmathematical subjects, purposes, and frames. Whole class discussions with the teacher were more mathematical due to the teacher's efforts to discuss the students' thinking, which shifted the focus away from answers to how the answers were obtained (McNair, 1994).

Students had engaged in several problem-solving activities similar to the one reported here and had had several whole class discussions concerning the identification and use of patterns in finding solutions. These discussions formed a corpus of experience and texts that could be used in the development of a mathematical context to discuss and solve similar problems. Data analysis revealed that all of the concepts needed to develop a formal representation of the equation of a line were implicit in the whole class discussions led by the teacher (McNair, 1994). Thus forming and working within an abstract context was within the limits of the texts created as a part of the students' experiences.

As a result of their cooperative group activities and the whole class discussions, there was no lack of mathematical texts on which the students could have built a mathematical context to support their problem-solving efforts. But, since they never adopted the mathematical frame that Sheila adopted, making the Type 2 problem just a prop for discussing mathematics, the relationship between their activities and these mathematical texts was not apparent or relevant to them. The following example will show how their failure to use and consult these texts was related to their nonmathematical discourse, so that when their problem-solving efforts stalled they were unable to create a structured mathematical context to assist their reasoning.

The students had been given data that indicated that a record manufacturer paid 30 cents less for blank CDs for every additional 250 CDs made, and they had been asked to determine how much it would cost to manufacture 3,674 CDs. This problem was similar to many of the problems the students had worked on in that there was an easily recognizable pattern that could be used to determine the cost of manufacturing CDs in increments of 250; however, this left an extra 174 CDs that could not be handled in this way. The intention was that these extra CDs would focus the students' attention on the mathematical relation between the cost of the blank CDs and the number produced. The dialogue reported below was recorded as the students began to deal with the extra 174 CDs.

68 Warren: And you're going up another one seventy-four
69 so you gotta figure a way to divide the thirty by one seventy-four to
70 see how many cents that equal up to one seventy-four.
71 Dave: So what we want to do is divide point three.
72 Warren: (. . .)
73 Dave: Divided by one seventy-four\\
74 Warren: \\So every two fifty records it's going up thirty cents right?
75 Dave: That's fourteen point two nine.
76 Warren: Ha?
77 Dave: Fourteen point two\\
78 Warren: \\No we're trying to find thc number.
79 Like one seventy-four would be left over so we have to find out what
80 the thirty cents is equivalent to in the one seventy-four.
81 (. . . 2 sec . . .)
82 Wesley: Da what?
83 One seventy-four?
84 I thought we got\\
85 Warren: \\Yea because . . . OK it goes up thirty cents.
86 OK, every two fifty records is thirty cents.
87 So two fifty times two is five hundred.
88 So you have one seventy-four left over.
89 So you gotta find out how many cents, OK, if it goes up
90 one seventy-four . . . one seventy-four copies
91 how much . . . how many cents will it go up by that?
92 Wesley: So (. . .) ah hundred seventy () right?
93 Warren: Ah . . . ha.
94 So you have like one seventy-four.
95 Dave: Divide () by thirty () more than that.
96 Because one seventy-four divided by thirty cents is fifty-eight cents, so that's\\
97 Warren: \\Yea
98 You have one seventy-four divided by thirty . . . and that would equal five point eight cents.
99 Dave: Five point eight cents?
100 Warren: Yea . . . or five dol . . . yea.
101 Wesley: Don't you mean five dollars and eighty cents.
102 Warren: No five point eight cents.
103 Wesley: (Don't give me no five point eight cents.)

104 Warren: So if you just round it off.
105 Dave: Just round it off to six.
106 Three twenty-six point three dollars and twenty-six cents.
107 Warren: Then after we do that do we do six times three thousand six hundred seventy-four?
108 Dave: Sounds good.
109 Charles: OK.

Warren's utterance in lines 68–70 was the tip of a mathematical text that could have led to a discussion of the constant proportional change between the variables in the problem. At this point, the stage was set for all of the previous whole class discussions and cooperative group activities to come together to support the students' efforts to solve this particular problem. There were a number of whole class discussions that were particularly relevant to the question Warren raised that could have been used to form a context for the discussion.

Warren had identified a mathematical subject; however, the group's purpose was to find an answer to the problem and not to develop a deeper understanding of the relationship between the variables. This nonmathematical purpose is a major reason the students did not create a mathematical text. In the two previous examples the mathematical context provided by the teacher insured that the students would have mathematical purposes by restricting their choices; however, in this last example it is a mathematical purpose that is needed to create a mathematical context.

In his statement of the problem in lines 68–70, Warren used the word *divide* to indicate that only a fraction of the 30-cent unit decrease would be made and to indicate that this fractional part was somehow determined by the number 174. Dave, working in a calculation frame, misinterpreted Warren's use of the word *divide* and took it as a suggested method to calculate the solution instead of as an attempt to understand a numeric relationship based on the concept of division. Dave's failure to offer an explanation or to justify the validity of what he was doing is further evidence he was not working in a mathematical frame.

The difference between Dave's and Warren's frame was evident in lines 73–78 as they talked past each other. In lines 78–80, Warren restated his question in an effort to get the other group members to adopt and contribute to the text he was attempting to build. This time his statement, "we have to find out what the thirty cents is equivalent to in the one seventy-four," was close to the form "174 is to 250 as 30 is to x." This was in fact the mathematical question he needed to solve; however, he was missing the relationship between 174 and 250. He understood that 174 was only a part of the 250, but he did not understand that it was that percentage (i.e., 174 divided by 250) that he needed to find his answer.

In the Pirie and Kieren study this problem was eliminated by the fact that the students had engaged in several activities that used the needed mathematical concepts under more controlled circumstances. Sheila's efforts to interpret the egg carton problem in a mathematical way were supported by the frame and the procedures that had been developed with the teacher. In Warren's case, the point of the activity was to develop the procedure, and to do this he needed to adopt a mathematical frame. As in many cases when mathematicians do mathematics, Warren and his peers could not rely on the familiarity of well-known procedures and concepts to find their answer. It must be emphasized again that developing the procedure they needed was well within the students' ability. It was not, however, part of their purpose or their frame.

When Wesley seemed confused but willing to entertain Warren's ideas, Warren restates the question again, in lines 85–91, in an effort to provide a context for their discussion, but again without success. Finally, in lines 95 and 96, Dave repeated his suggestion to divide, and perhaps because his question depended on a fractional relationship between 174 and 30, Warren seemed willing to accept this as a possible solution. The only justification offered, however, was the result of the calculation in line 96, which again reflects the fact that the group was not working in a mathematical frame.

At this point in the discussion, even Warren abandoned the mathematics frame as the group attempted to place the decimal point in their quotient. Since 5.8 cents did not make sense to the students, they suggested 58 cents or $5.80 and finally decided to round off to 6 cents. This was not a mathematical decision or discussion. The students used their contextual knowledge of money to place the decimal point. From a record manufacturer's perspective this may be a perfectly logical thing to do, but from a mathematical perspective this was an inappropriate shift in frame.

In the end, although the building blocks to construct a mathematical text existed within the corpus of the students' prior activities and experiences, due to the nonmathematical discourse and, specifically, the nonmathematical frame, the students did not build a mathematical context to help with their problem-solving activities. Unlike in the previous two examples, there was no mathematical context provided by the teacher to restrict the students' discourse choices, and the students were not able or inclined to produce one. Warren's efforts to create a mathematical context, however, are not to be minimized. Although he was ultimately unsuccessful, this is exactly the kind of thing students must learn to do on their own. If the group had accepted his question and shifted their focus from finding answers, the results of the discussion would have been significantly different.

It may be argued that the students' focus on finding answers was appropriate due to the nature of the problems they were asked to solve. During the early stages of the problem-solving activity the students had appropriately based their utterances in the problem and calculation frame as they engaged in the process of understanding the problem by identifying the key variables and finding patterns. However, when confronted with Warren's mathematical problem the students did not collect terms and definitions to establish a mathematical context for discussing the problem. They did not refer to their previous problem-solving experiences or the problem-solving rules that they had developed as a result of previous problem-solving activities and whole class discussions. Instead they continued to try to find answers by applying simple calculations based on their intuitive interpretations of the problem context. They were not working as mathematicians in a mathematical frame, they were working as producers in a real-world problem frame.

Another important observation concerning the frame that supported the students' utterances is the lack of a peer frame, a frame that existed in both the Lampert and the Pirie and Kieren studies. The development and use of a peer frame is essential to the development of a common frame shared by all of the group members. In the previous examples the well-developed peer frame was again supported by the mathematical context that served as a common referent for all of the students. In the Pirie and Kieren study this is clearly evident as Sheila gets the other group members to adopt her frame, which is the frame introduced by the teacher along with the mathematical context. In addition, in the Lampert study the development of communication skills to support the use of a peer frame was a specific focus of instruction.

Discussion

In mathematics, discourse is used to create a text that forms a context for discussion and knowledge construction. It is clear from these three examples that the context provided by the teacher plays an important role in the development and use of mathematics discourse and that discourse plays an equally large role in the development of a mathematical text. The first two examples demonstrate that students can work autonomously in, and with, mathematical contexts provided by the teacher, largely due to the restrictions imposed on their discourse by those contexts. The last example suggests that students with poor discourse skills may have some difficulty creating their own mathematical contexts.

The frame used to interpret a situation will determine which tools and solution processes are appropriate. If the frame is mathematical, then mathematical tools and solution processes will be appropriate. If students are to develop the ability to analyze a situation – Type 1 or Type 2 – to determine what is appropriate, they will need to develop mathematical frames to guide their efforts. To do this students may need the kind of controlled experiences provided by the Lampert and the Pirie and Kieren examples, but they must eventually grow to be able to develop their own mathematical context where one did not exist. Students will need to learn to make the shift (Cobb, Gravemeijer, Yackel, McClain, & Whitenack, in press) that Sheila made and to develop abstract contexts like Lampert's students did, but without the support of an initial prototype to work in and build on.

The teacher's role as the most highly trained mathematician in the group is not to dictate and dominate discussions or to provide students with answers. The teacher's role is to help the students form a productive mathematical community and to provide quality control for the products produced by the community. Teachers may do this by asking questions that provide opportunities for students to develop their discourse skills under more or less scaffolded conditions. In the first example the teacher scaffolded activity by asking the students questions about abstract numeric relationships and allowing and encouraging them to explore their own understandings. In the second example the teacher scaffolded the students' activities by providing practice with the concepts the students would need to solve the egg carton problem. While there was a conscious effort by the teacher to minimize scaffolding in the third example, the choice of problems that were related by the same abstract mathematical concept and the whole class discussions provided a kind of indirect scaffolding.

Teachers are responsible for making decisions every day that shape the classroom culture in which students work and learn and that provide the students with opportunities to do some things and not others. The decisions teachers make will determine the opportunities students have to develop a mathematics discourse. One idea that may guide these decisions is that students should be engaged in authentic mathematical activities that provide expanded discussion opportunities and that require students to construct their own contexts to answer questions.

The students in the first two examples did not have to use a set of mathematical beliefs to construct a context since it was provided by the teacher, and their motivation did not have to be mathematical for their activities and discourse to appear to be mathematical. This does not imply that their beliefs and motivation were not mathematical, but that the problem context did not require them to be. The students in the last example were unable, or at least not inclined, to apply a mathematical perspective and so their subject, purpose, and frame remained nonmathematical. With the exception of Warren's efforts to establish a mathematical context, they did not act as if guided by the beliefs, values, and expectations of the mathematics community.

Frames shape our impression of a situation and allow us to interpret the talk associated with the situation (Goffman, 1975; Tannen, 1993). In mathematics we must purposefully choose to use a frame that may not be a typical choice for a given situation, as Sheila did. Students will need to have well-developed mathematics frames in order to shape their interpretation of contexts in a mathematical way. This is what the teachers' scaffolding did for the students in the first two examples and what students must be able to do for themselves. This is a major part of the problem formation and solution process and so students must be able to do it on their own.

The major difference between Type 1 and Type 2 problems is that in the real-world context the student must create a world of realities that allows him or her to make declarative statements about the mathematical realities of the situation. This requires that the situations be abstracted in a particular way. In the abstract case of a Type 1 problem, the world of realities is defined by the teacher. Whatever meanings students are able to construct are based in this abstract, often syntactical, world of numbers, symbols, and patterns that have no external referents. Learning to work in such contexts may not prepare the student to deal with the problems encountered by Warren and his peers when working in the real-world context of a Type 2 problem.

As suggested earlier in the chapter, it is unlikely that instruction based exclusively on one problem type will be as productive as instruction based on a mixture of the two. What is not clear is how to move the students and instruction from the first example, where mathematical discussions are facilitated by the abstract mathematical context provided by the teacher, to the second example, where the students' discourse is facilitated by isomorphic relationships between problems and instruction, to the last example, where students must create their own problem-solving context with minimal support. It is also not clear that this is the proper or only order. The reverse order makes sense as does starting with the situation represented in the second example.

Conclusion

Due to the abstract nature of mathematical discussion, the context that supports it may have little connection to real-world concrete situations, so that the semantic content of utterances must be abstracted from the texts of previous utterances and discussions. Constructing these texts plays an integral part in the development of mathematical culture and discourse, and so it makes sense that their construction should play an important part in the development of mathematical culture and discourse in the classroom.

Longer studies are needed to determine the effects of problem type and teacher scaffolding on students' mathematics discourse development. The results of the analysis here have identified some of the processes that need to be addressed by future research. Problem type was used here to compare and contrast the three examples; however, there are also significant differences in the instruction, content, and the ages of the students in the examples. Each of these variables alone and in combination may affect the development of mathematics discourse. The examples from the Lampert and the Pirie and Kieren studies demonstrate how instruction can interact with problem type to increase discourse development activities. What is clear from this discussion is that a consideration of the discourse opportunities provided by a problem should contribute to the instructional choices made by teachers.

Acknowledgments

I would like to acknowledge the contributions of Paul Cobb, Deborah Hicks, James Hiebert, and Magdalene Lampert for their help in shaping the ideas in this chapter and for providing comments on earlier versions.

Notes

1. In the following transcript excerpts, cited from Lampert (1988), numbers are written as numerals as they were in the original text. In transcripts original to this volume, we adhere to the convention of spelling out numerals in dialogue.

References

Brown, G., & Yule, G. (1983). *Discourse analysis.* New York: Cambridge University Press.

Cobb, P., Gravemeijer, K., Yackel, E., McClain, K., & Whitenack, J. (in press). Mathematizing and symbolizing: The emergence of chains of signification in one first-grade classroom. In D. Kirshner & J. A. Whitson (Eds.), *Situated cognition theory: Social, semiotic, and neurological perspectives.* Hillsdale, NJ: Erlbaum.

Cognitive and Technology Group at Vanderbilt. (1993). Anchored instruction and situated cognition revisited. *Educational Technology* (52–70).

Goffman, E. (1975). *Frame analysis.* New York: Harper and Row.

Lampert, M. (1988). *The teacher's role in reinventing the meaning of mathematical knowing in the classroom* (Research Series No. 186). East Lansing: Michigan State University. Institute for Research on Teaching.

McNair, R. E. (1994). *A descriptive analysis of students' mathematics discourse in cooperative group and whole class discussion.* Unpublished doctoral dissertation, University of Delaware, Department of Education, Newark.

McNair, R. E. (1996, April). *A frame analysis of mathematics classroom discourse: What are students doing with discourse?* Paper presented at the annual meeting of the American Educational Research Association, New York.

Pirie, S. E. B. (1991). Peer discussion in the context of mathematical problem solving. In K. Durkin & D. Shire (Eds.), *Language in mathematical education: Research and practice,* pp. 143-161. Philadelphia: Open University Press.

Pirie, S. E. B., & Kieren, T. E. (1992). Creating constructivist environments and constructing creative mathematics. *Educational Studies in Mathematics, 25,* 505–528.

Schoenfeld, A. H. (1992). Learning to think mathematically: Problem solving, metacognition, and sense making in mathematics. In D. A. Grouws (Ed.), *Handbook of research on mathematics teaching and learning,* pp. 334–370. Reston, VA: National Council of Teachers of Mathematics.

Snapper, E. (1979). What is mathematics? *The American Mathematical Monthly, 86,* 551–557.

Tannen, D. (1993). What's in a frame: Surface evidence of underlying expectations. In R. Freedle (Ed.), *New directions in discourse processing,* pp. 137–181. Norwood, NJ: Ablex.

5 Disciplined Perception: Learning to See in Technoscience

Reed Stevens and Rogers Hall

> It is futile to study perception "in itself." It must be treated as a "phase of action" in relation to the motor and intellectual activity of the individual. . . . An object only affects behavior in so far as it has meaning, and this only arises from its functional relations to other objects, be they spatial or temporal relations, or relations of causality or purposiveness, etc. The problem of meaning, therefore, ultimately has priority over that of form.
>
> Albert Michotte (1954)

> Let the use *teach* you the meaning.
>
> Ludwig Wittgenstein (1953)

Introduction

This chapter presents an approach to understanding how people learn and use their bodies to participate in the cultural practices of technoscience.[1] Our analyses are organized around two detailed cases of embodied disciplinary knowledge drawn from two naturally occurring settings. The first case is a tutoring interaction involving an adult tutor and an adolescent student working together on school mathematics tasks involving the Cartesian coordinate system. The second case involves two civil engineers redesigning a roadway plan for a housing project using a more complex "coordinate system." These case studies describe the development of what we call "disciplined perception."

This volume attests to the growing focus on talk in research on mathematics learning. As conversation and discourse analysts have argued over the last twenty years, "talk-in-interaction" (Schegloff, 1992) is a primary site for activity that makes, reproduces, and transforms our social and cognitive worlds. However, as a decade of studies also attest (Goodwin, 1995; Goodwin & Goodwin, 1996; Goodwin, 1994; Latour & Woolgar, 1986; Lynch, 1985; Lynch, 1990; Suchman, 1987; Traweek, 1988), technoscientific practices include a great deal more than talk. In particular, these studies highlight the thoroughgoing materiality of technoscience. If a focus on talk makes interaction between people central, then a focus on materiality makes interaction between people and technoscientific things (texts, inscriptions, machines) central. In the case studies that follow, we analyze both these forms of interaction, because our ongoing fieldwork repeatedly reminds us that the "situated activity systems" (Goodwin & Goodwin, 1996) we find in offices, classrooms, science museums, and laboratories are intricate webs of interaction involving both people and things.

One way technoscientists interact with things, especially with inscriptions like graphs, diagrams, and models, is visually. As Marx Wartofsky said in describing scientific activity, "we see by way of our pictures" (Wartofsky, 1979). In this chapter, we explore the question of how people learn to see with and through their inscriptions. Our theoretical position will be that forms of visual interaction develop and stabilize – what we call "disciplined perception" – in interactions between people and things. Our case studies show that in these interactions, people demonstrate to each other the relevancies that organize their work, and it is these relevancies that shape specific forms of visual interaction.

The term *discipline* is used to mean two different but related things. First, we assume that contemporary technoscience, in and out of school, has a disciplinary organization (Messer-Davidow, Shumway, & Sylvan, 1993). Just as specific disciplines have characteristic modes of thought, genres of verbal interaction, and styles of dress, they also have characteristic visual practices. If someone can sound like a doctor, he or she can also see like one, visually and physically inspecting bodies for signs that meet doctors' criteria of "illness" or "abnormality." If someone can look like an architect, he or she can also look at things like one. This sense of discipline demands that we carefully describe visual practices, both in relation to the tasks, artifacts, and settings where they are deployed and in relation to other embodied practices (e.g., pointing and gesturing) that support them. We think of disciplined perception as a performance genre – a set of specific forms of embodied action. Accordingly, cognition is a property of the entire body in action, a body shaped and maintained by participation in disciplinary (and other) communities.

Our second sense of discipline complements the first. Because forms of perception differ across disciplines, then within disciplinary communities we should expect to find interactional and organizational means through which disciplined perception is learned. This sense of discipline roughly means "instruction and exercise designed to train to proper conduct or action" (*Random House*, definition #2). However, we assume that learning to participate in disciplinary practices does not depend solely on "instruction and exercise" as it appears in Western schooling, but instead can occur in varied forms of apprenticeship (Becker, 1986; Lave & Wenger, 1991; Rogoff, 1990). In this chapter, our approach is to describe indigenous forms of learning and teaching as we find them, whether they are explicit, tacit, verbal, kinesthetic, and so forth. The word *discipline* (rather than teaching, training, or instruction) also marks an important moral quality (i.e., regarding "proper conduct or action"). In the lived social world, just as ways of talking or dressing occasion judgments and acts of discipline by others, so too do ways of seeing. In terms of disciplined perception, this moral aspect of practice appears interactionally in evaluations of how members see and the implications they draw from what they see, but also in evaluations of what they treat as properly visible and invisible.

Disciplined Perception and Coordination

A familiar example of disciplined perception involves the capacity of experts to quickly register perceptual features that are relevant to their particular practice, features invisible at a glance to nonexperts. Accounts of this celebrated capacity circulate in both academic (deGroot, 1965) and informal discourse (e.g., chefs tasting ingredients, wine experts determining a vintage, sociolinguists hearing dialects, etc.), but we use *disciplined perception* to mean a broader set of actions that have an extended duration in time and frequently stretch across people and things. The naturally occurring events we document are sequences in which people assemble and *coordinate aspects* of visual displays[2] to make practically relevant objects or conditions visible to themselves and coparticipants.

We have chosen to use the word *aspect* rather than the more familiar *feature* to avoid the naive realism that often leads cognitive researchers to attribute features to visual displays, as if these features were *there-for-all-to-see.* In our view, studies of practical perception should not theoretically assign the status of feature to the outcome of a visual activity; the stability and obviousness of a feature in practical visual experience is an empirical question. Our opposition to features-as-given types of analysis is similar in kind to what Street (1984) calls an autonomous model of literacy, in which the mere presence of certain forms (e.g., textual literacy) determines which kinds of cognitive activity are found. At least with regard to visual practices in technoscience, we treat the seeing of aspects as a practically and contingently accomplished relation between viewer, material display, and practical situation.[3]

The meaning of *coordination* that we use builds on work by Hutchins (1995), who describes the systematic coordination of actions and media as a necessary achievement in the distributed work of Navy navigation. We also borrow from diSessa, who describes cognitive categories like quantity or physical object in terms of "coordination classes" (diSessa, 1991) that integrate action, perception, and representational forms. For Hutchins, a Navy ship finds its way on the open ocean by repeatedly achieving states of coordination across people and representational devices (e.g., charts, alidades, and differently located shipboard personnel). In Hutchins's account, people and devices produce "representational states" that are "propagated" through the navigation cycle. In our use of coordination, we share Hutchins's focus on distributed activity across people and representational media, but we foreground the active, embodied practices by which people bring these heterogeneous elements into coordination. We focus less on coordination via the propagation of distinct states and more on coordination between finer-grained actions that *are assembled to comprise these distinct states.* We argue that when learning is a central concern, the work of comprising states can never be taken for granted. DiSessa's account also attends to fine-grained phenomena as when physicists observe quantum wave functions, through "indirect . . . radically non-everyday coordinations." We agree with diSessa on the scale of these phenomena (i.e., assembling states) and centrality of learning, but we do not adopt his units of analysis which are internal cognitive processes. Like Hutchins and others (Goodwin, 1995; Goodwin & Goodwin, 1996; Goodwin, 1994; Latour, 1990; Lynch, 1990), we are concerned with the materially mediated, distributed character of real world perception. Our case studies examine how the forms of coordination we call disciplined perception are learned, used, and maintained in social practice.

Comparing Cases of Disciplined Perception

This chapter compares two cases of activity in which people use coordinate systems of representation to enable joint work. In each case, there are differences in the power and knowledge of people who work together. Our analysis focuses on two questions: (1) How do participants in school and work situations demonstrate to each other a shared understanding of some specific coordinate system of representation? (2) How do breakdowns in these shared understandings lead to significant teaching and learning events across these different situations? Both questions are central for understanding how people use language together when engaged in mathematical practices, linking our work to the collective theme of this edited volume.

In our first case (tutoring), the tutor and student do not share a mathematically disciplined perception of the Cartesian plane and its relation to mathematical tasks. As a result, their different uses of this coordinate system have significant and observable consequences for joint activity. Under these circumstances, breakdowns in shared meaning are frequent, so we have ample opportunity to examine breakdowns and the detailed interactional work of disciplining perception. In the second case (civil engineering), a junior and a senior engineer mostly share a disciplined perception of roadway design, and this facilitates their joint activity in ways that are unavailable in the tutoring situation. Under these circumstances, breakdowns in shared meaning are rare, so we can examine how relatively stable forms of disciplined perception are used and maintained. However, even among these civil engineers, breakdowns do occur and lead to significant teaching and learning events (Hall & Stevens, 1996). This allows us to compare the interactional organization of disciplining perception across cases.

Our first case demonstrates how the practical terms *figuring it out* and *checking* are grounded differently in activity for the tutor and the student. The student grounds these terms in visual activities with what we call a grid view, while the tutor grounds these terms in coordinated use of representational materials (e.g., graphs, equations, and ordered pairs) that readers will recognize as conventional mathematics. Much of our analysis focuses on what the tutor and the student do so that their differing visual practices begin to converge over the course of the tutoring sessions. Our second case demonstrates how "looking" and "seeing" are complex, coordinated performances that are made possible by a more stably grounded set of visual practices. Much of our case analysis documents this coordinated form of disciplined perception, but we also examine what happens when these coordinated meanings break down.

Before proceeding with the case studies, we present some general terms that we will use when analyzing both cases. We use the term *view* to refer to the major forms of inscription that circulate in each setting. The term is borrowed from the engineers, who work with three views of roadways: plan, station, and profile. In the tutoring case, we will focus on two inscribed views, the first a 20 by 20 grid of dots with Cartesian axes centered in the middle (the grid view) and the second showing these axes without the grid (the blank view). Within the tutoring sessions these two renderings of the familiar Cartesian coordinate axes support distinctly different ways of seeing mathematics. We also use the term *visual orientation* (Coulter & Parsons, 1990) to describe the variety of ways in which people attend visually to the material environment around them. More specific terms are introduced to the extent that we can demonstrate their relevance to the participants in the case studies. This distinction allows us to pay attention to important differences in the ways participants use verbal language to refer to their visual activities and to track how these uses change and often converge over time.

Case 1: The Tutoring Interaction: Learning to See in Cartesian Space

Background

The first case documents the work of an eighth-grade student (Adam) and an adult tutor (Bluma) engaged in six tutoring sessions as part of a university study of mathematics learning. Approximately 6 hours of audio/videotape (about 1 hour for each of six tutoring sessions) were recorded by the Functions Group (Alan Schoenfeld, his students, and affiliated scholars at the University of California, Berkeley) as part of a tutoring experiment. The study was designed to document mathematics learning in a number of tutoring interactions and to investigate how a computer tool could contribute to learning in the topic area of linear functions and their graphs. The tutoring sessions took place in a small room, furnished with two chairs and a computer on a table. Cameras recorded from two perspectives – from above and from behind – thus capturing work on the flat table surface in front of the computer as well as work on the computer screen. Written documents of all work were also collected.

The tutor and student worked through a sequence of tasks. Some of these involved calculations or graphing on paper, while others directed the student to use the computer to accomplish the mathematical tasks. The first tutoring session involved only work on paper, and this session was designed primarily to be a review. In this session, the student was asked to plot points in the Cartesian plane and to become familiar with the language of "coordinates," "quadrants," "lines," and so forth. Most of the work described in this case comes from sessions two through five;

in these sessions the student was directed to explore the relationships between linear equations (i.e., of the form $y = mx + b$, where m represents the slope of the line and b represents the y-intercept) and their graphs in the Cartesian plane. Some of the tasks were exclusively about the effects of changes in either the slope or intercept parameter, while other tasks involved changes in both parameters.

At the time of the study, the student was enrolled in a pre-algebra class. He had experience plotting points on the Cartesian plane, but most of the curricular activities that he encountered during the six videotaped sessions were unfamiliar to him. The tutor, holding a doctorate in mathematics, was quite familiar with graphing functions on the Cartesian plane and was an experienced tutor. Some postsession interviews were conducted between the tutor (Bluma) and a member of the research staff, and one interview was conducted with the student (Adam) at the conclusion of the sessions. In the interviews, Bluma described herself as trying to take a "nondirective" and "laissez-faire" approach. The data reflects this approach in that little straightforward instruction is observed. Instead, Bluma repeatedly encourages Adam to formulate his own methods for accomplishing the tasks and repeatedly insists that he answer his own questions.

Analysis

Our case includes eight episodes, selected from materials collected and analyzed over the 6-week tutoring experiment. We have selected episodes from these weeks to illustrate the development, use, and eventual transformation of what we call a "grid calculus" on the part of the learner (Adam). By attributing a grid calculus to Adam, we mean that he has a coordinated set of embodied practices for constructing and answering questions about graphs of linear functions within a *specific* representation of the Cartesian plane, the grid view. The grid view is presented early during the teaching experiment, and it contains a variety of material supports for drawing lines and making calculations that are important in Adam's work. In overview, our analysis of the case shows four things: (1) visual practices that make use of aspects of the grid view are the primary means through which the student "figures out" the tasks; (2) the deletion of parts of the grid view creates unexpected breakdowns as the learner and tutor progress through a sequence of tasks; (3) much of the tutor's work in "disciplining" the attention and activity of the learner is triggered by these breakdowns; and (4) the learner's understanding of

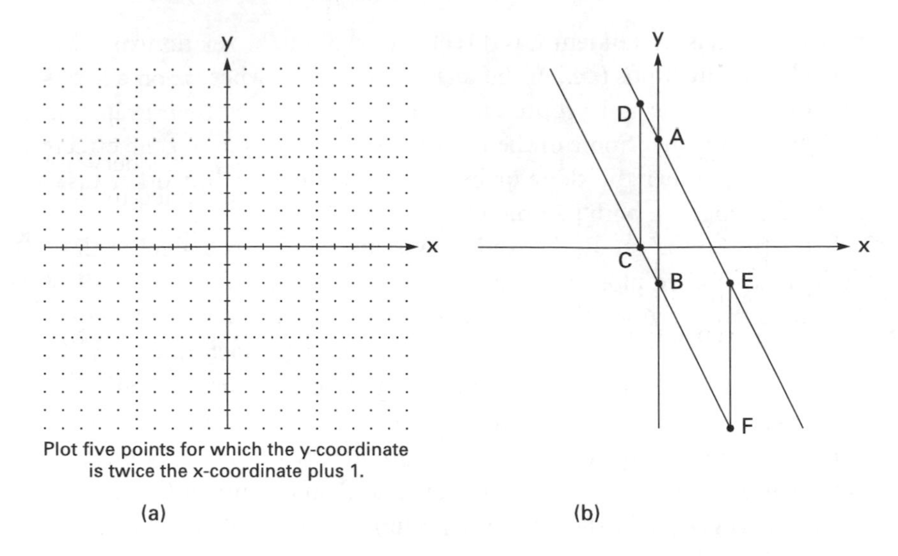

Figure 5.1. Two views of Cartesian space. Figure 5.1(a) shows the grid view and the text from one of the tasks the student encountered early in the sessions. Figure 5.1(b) shows the blank view with drawn lines and labeled points that are part of a task the student encountered late in the sessions (episode 7).

linear functions on the Cartesian plane is eventually reorganized in response to both the alteration of the material environment and the tutor's acts of disciplining Adam's perception.

Figure 5.1 shows representative task environments presented to Adam during the first and last weeks of the tutoring experiment. Part (a) shows an example of what we have called the grid view, while part (b) shows an example of the blank view. In the grid view, three conventions support Adam's use of a grid calculus:

- Horizontal and vertical axes are partitioned with tick marks at unit increments – e.g., points (1, 0) and (2, 0).
- A grid of points fills bounded regions in all four quadrants of the plane, with a finer series of points projecting out at intervals of five units.

The bounded region in each quadrant has visually prominent "edges" where the grid of points ends.

As will be clear in the episodes, these supports (i.e., tick marks, a uniform grid of points, and the edges of this grid in each quadrant) form what Adam evidently takes to be a well-ordered medium for drawing lines, making judgments about the graphs of equations, and counting point or segment values. Again, we will call this specific representational form the grid view. To utilize the grid view, Adam draws upon a set of presumably simple visual capacities:

- the capacity to notice and count various displacements between objects in the grid view as being equivalent or nonequivalent (e.g., between parallel lines, between the origin and the x- and y-intercepts, and between lines and grid corner points where graphed lines have rotational symmetry).
- the capacity to produce a straight line by following points in the grid view (e.g., to draw the line $y = x$ by drawing or tracing from (10, 10) to (9, 9) to (8, 8), and so on).

The grid view is a particular way to represent the Cartesian plane, and as the tutoring sessions proceed, this representation changes. By week five, as shown in part (b) of Figure 5.1, tasks begin to ask for inferences about equations and the determination of intercepts using a Cartesian plane without partitioned axes or a grid of points. While we and the tutor (Bluma) understand that ordered pairs, equations, and graphs of linear functions can be drawn and systematically understood as the same across these two views – an understanding supported by having what Schoenfeld, Smith, and Arcavi (1993) call the "Cartesian connection" – our analysis of Adam's responses to these and other tasks shows that, for him, this normative, coordinated understanding of functions is still under development.

We now consider these issues in a series of eight episodes, selected from the 6-week tutoring experiment. In the transcript segments that follow, successive turns-at-talk are rendered along with descriptions of accompanying action. Numbers within the transcribed talk indicate the onset of an action described below the turn. Within the transcribed talk, we use "=" to indicate latching and "/" to indicate overlapping talk. CAPS indicate audible emphasis. Our transcripts also give embodied action nearly equal prominence as talk in order to show the extent to which action and talk are co-articulating resources in mathematical speech exchange situations.[4] Representing talk and action together allows us to demonstrate how and when participants make sense of each other's talk through interpretation of concurrent action and make sense of each other's action through interpretation of concurrent talk.

Episode 1 (Week 1, Tasks 1 and 9). During the first task of the first session in this tutoring experiment, Adam noticed "tiny dots" in the grid view and, when asked by the tutor about them, reported that they were for counting. As he whispered through count sequences to find the values of ordered pairs, the tutor explicitly told him, "You can count out loud," and he expressed surprise. From the beginning of this teaching experiment, then, the grid of points was a salient, orienting resource for Adam.

On the ninth task during this first session (shown in Figure 5.1[a]), much of what we call Adam's grid calculus was already strongly in use. After counting up and over to locate points for several ordered pairs, Adam made a freehand drawing of a line passing through these points, and Bluma asked him to "figure out" the intercepts.

1 Bluma: So, you know, can you figure out exactly where it hits this x-axis here (1)?
1 (points at line Adam has drawn, near x-axis)
2 Adam: Well, I probably didn't draw it (1) so straight, but, I guess =
1 (pen traces along drawn line toward the x-axis)
3 Bluma: = I don't know, it looks pretty good to me, do you know =
4 Adam: = Yeah, I =
5 Bluma: = But can you figure it out anyway?
6 Adam: (1) Yeah, uh, negative a half . . . zero.
1 (moves pen to point of intersection with x-axis)
7 Bluma: Sounds reasonable to me (1).
1 (nodding affirmatively)
8 Adam: (1) Alright.
1 (nodding affirmatively)
9 Bluma: Um, and so this thing here (1) that goes through the y axis is? Is that um, so you've drawn it at =
1 (moves pencil to point of intersection with y-axis)
10 Adam: = zero, one =
11 Bluma: = one, zero (laughing) zero, one. Right.
12 Adam: OK. (clipped tone may suggest he is ready to move on)

Asked to "figure out exactly where" a line crossed the x-axis, Adam was aware that precision was important (e.g., "I probably didn't draw it so straight"), but he still traced along the line to find an answer by visual inspection. After quickly providing the y-intercept, he appeared ready to move on. But the tutor persisted, asking whether line tracing and visual inspection were backed by some other means of figuring out the intercepts. As they continued, she questioned whether he had "checked it in [his] head," presumably calling for mental calculation to find the y-intercept (e.g., solving $0 = 2x + 1$).

Asked about checking his solution to what he apparently took as a simple problem, Adam replayed his manual/visual inspection of the drawn line, then dismissed the possibility that the line might "veer" into another quadrant and pointedly asked if the tutor was ready to "move on."

In our view, this first episode shows several things about Adam's understanding and the character of his interaction with the tutor. First, substantive parts of a grid calculus are clearly available to Adam: (a) he attends to the grid of "tiny dots"; (b) these comprise a uniform medium for oral/manual counting with the grid view; and (c) he knows that lines drawn between points on the grid are imperfect but they are still good enough for locating points of intersection and finding their values. Second, and without explicitly disqualifying these manual/visual activities, the tutor presses for different mathematical activities with phrases like "figure it out exactly" and "check it in your head." Third, Adam appears to resist these attempts to discipline his activity, because he is confident with his own means of "figuring it out"; he even pushes the tutor to get on with the next task. In sum, we think this episode shows a learner with a stable approach to tasks within a specific representational environment. As part of the tutoring experiment, however, the tasks and the environment will change, and this takes us to our next episode.

Episode 2 (Week 2, Task 2). In this episode, Adam is given a task sheet showing a graphical "starburst" (see Figure 5.2[a]) and asked to reproduce this image using computer software. The computer software allows a user to type in equations that are automatically graphed in a window with a grid view, and these graphs accumulate on top of the grid as more equations are entered by the user.

Prior to the activity we transcribe in this episode, Adam has produced a number of graphs that pass through the upper half of the first quadrant (i.e., lines with slopes 1, 2, and 3 in Figure 5.2[b]). As our transcript begins, Adam asks Bluma if he should "just keep doing this." This question leads Bluma to orient Adam's visual attention to aspects available because of the layering of the starburst lines on the grid view; she points out that lines are "sort of evenly spaced" along the upper edge of the starburst picture. Once Adam is oriented to even spacing and he graphs another line, he assesses his efforts negatively ("I'm not doing it right"). Instead of evenly spaced lines he has produced lines that are "getting closer together."

1 Adam: So, I'm just going to keep doing this? (1)
  1 (hands on keyboard to type in another equation)
2 Bluma: Well, so you're trying to get this picture. (1) Alright?
  1 (holds up paper with starburst picture)

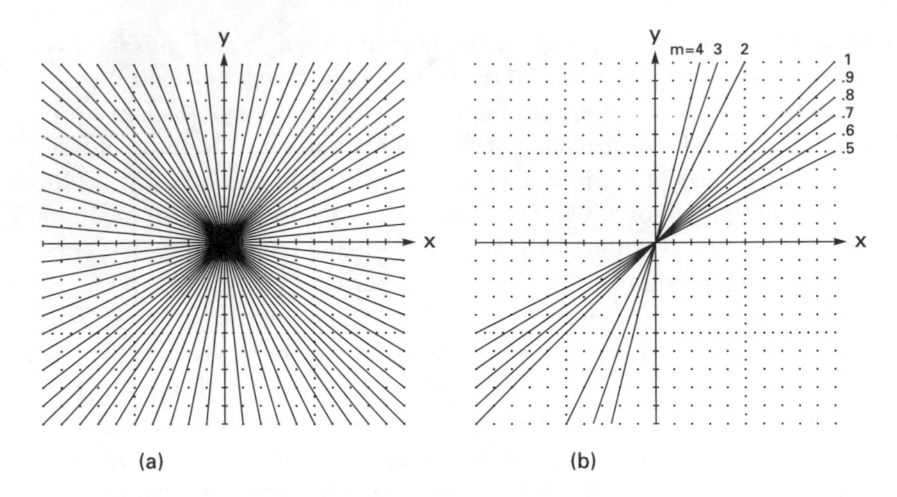

Figure 5.2. Starburst. Figure 5.2(a) is the image Adam was asked to reproduce. Figure 5.2(b) shows some of the graphs Adam generated while working on the starburst task in episode 2. Alongside the lines are the actual slopes of these lines to indicate that a linear relationship is correctly inferred along the vertical edge of the grid but not along the horizontal edge.

3 Adam: Well it looks like I'm getting it.
4 Bluma: Uh-huh, OK, so one of the yeah, so, so in some sense these lines (1) are sort of evenly spaced
1 (points to upper edge of the starburst picture)
5 Adam: Well I guess if I keep on (inaudible) the graph. I guess I can just keep on like this. (1) What? (3 sec.) OK, I don't think I'm doing it right (2 sec.) cuz they're getting closer together.
1 (types in equation $y = 4x$ and watches the computer graph it)
6 Bluma: Yeah, that's right, they are getting closer together.

At the end of this brief second episode, several things have happened that are important for our analysis of changes in this student's grid calculus. First, and as a result of Bluma's attempts to discipline his visual attention (turn 4), Adam begins attending to and using what we will call the "edges" produced by the array of grid points in each quadrant of the grid view. Although he initially uses visual comparisons along the horizontal edge of quadrant one to reject one approach to reproducing the starburst (i.e., slopes of 1, 2, 3, and so on at turn 5), Adam later clears the computer display and produces a family of lines with even spacing along the vertical edge of quadrant one (i.e., slopes of 1, .9, .8, .7, and so on, shown in Figure 5.2[b] and completed after the transcript shown above). He

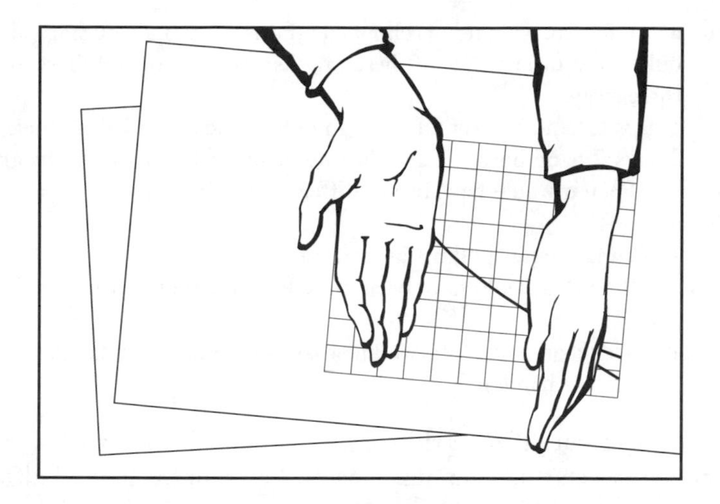

Figure 5.3. Tutor blocking view. The tutor used her hands to deny the student visual access to the grid dots along the upper and lower edges of the grid view.

judges this effort to be a success. Our second point, then, is that horizontal and vertical edges of the quadrants have become a durable aspect of Adam's grid calculus, and we should expect to see them used as a resource in subsequent tasks. As we show in the next episode, this is indeed the case.

Episode 3 (Week 2, Task 6). In this episode, with a grid view and the reference line $y = x$ on paper, the task directs Adam to "start with the equation $y = x$. Change it to the equation $y = 3x$. Predict how the graph would change." The task directs Adam to check the prediction with the computer. While working on this task, Adam incorporates the edges of the grid more fully into his grid calculus. Faced with a computer-generated graph that is clearly not "three away" from the line $y = x$, Adam remains oriented to the upper edge of the grid.

At this point, Adam is again working hard to coordinate the horizontal displacement along the upper edge of the grid view with changes in the slope parameter (i.e., equations of the form $y = mx$, where m is the slope). In one of her most direct interventions during these sessions, Bluma recognizes Adam's current reliance on the edges of the grid view and attempts to block these tactics by cutting off his visual access to these aspects (Figure 5.3 and turn 13, below).[5]

13 Bluma: Uh-huh, what if I (1) chopped off this paper and you start, and you only saw a little bit (2) of these graphs? Would you still have guessed three away?
1 (lays L hand horizontal in the middle of the upper half of the grid)
2 (lays R hand horizontal in the middle of the lower half of the grid)
14 Adam: Then it would have been a different graph.
15 Bluma: [(1)]
1 (Bluma's hands pull up, as if startled)
16 Adam: [Then] it would have been more like you know if you just got that part (1)
1 (traces segments of drawn line that fit between boundaries she made with her hands)

With the edges cut off the grid view, Adam makes what strikes his tutor (and us) as a startling announcement: The result will be a "different graph." It appears his understanding of functions and their graphs is so entwined with the aspects of the grid view, that by eliminating use of these aspects from the picture, the very objects being represented are also changed (i.e., "a different graph"). In contrast Bluma appears to believe that her actions only alter the parts of the graph that are visible (i.e., "only saw a little bit") and that as objects, they will maintain their identities (i.e., "these graphs"). While Bluma does not again block Adam's access to the edges of the grid view, she does continue to discipline his perception, showing him "other ways of looking at" displacements in the grid view. These actions, along with Adam's capacity to see symmetries and equivalent displacements between lines in the grid view, will be the resources for further development of his perception in subsequent episodes.

Episode 4 (Week 4, Task 2). If Adam's tactic of counting horizontal displacements for slope problems appeared shaky and seriously challenged in the previous task (episode 3), it is powerfully at work again during weeks 3 and 4. How could this be? Our answer is that the task environment becomes a more hospitable place for the grid calculus. Instead of asking Adam to predict how the line $y = x$ would change if the slope parameter is changed (as in episode 3), many tasks during these weeks direct him to predict how the line $y = x$ would change if the intercept parameter is changed (e.g., to $y = x + 3$ in this episode.) No further tasks ask him to relate changes of the line $y = x$ to changes in the slope parameter. In these hospitable circumstances, the grid calculus works. Figure 5.4(a) illustrates that $y = x$ is indeed "3 away" from $y = x + 3$ along the horizontal edge of the grid calculus.

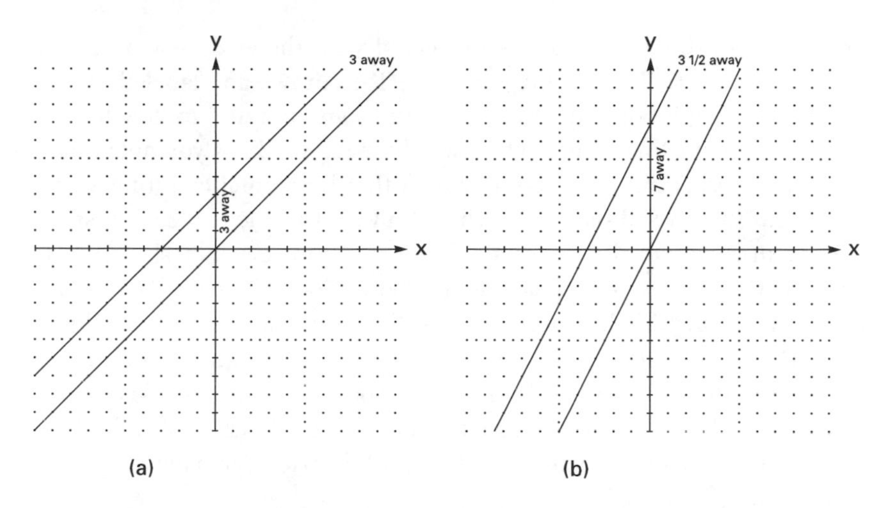

Figure 5.4. Task environments that are more and less hospitable for the grid calculus. Figure 5.4(a) shows the equivalence of horizontal and vertical displacements between $y = x$ and $y = x + 3$. Figure 5.4(b) shows the nonequivalence of horizontal and vertical displacements between lines when the slope is not equal to 1. In this figure, the displayed lines have equations $y = 2x$ and $y = 2x + 7$, a task environment Adam encountered in episode 5.

This episode presents a typical use of the fully elaborated grid calculus with a task of this kind. Adam gives a masterful demonstration of the grid calculus, exclaiming "Oh, that's so easy" and making quick use of the upper edge of the grid view. He establishes visual equivalencies to show that the reference line $y = x$ and the line $y = x + 3$ are horizontally 3 away everywhere in the grid view, and he even demonstrates his method for an equation that is not on the task sheet. Adam has become articulate about "the way [he] can figure it out," the resources he needs to introduce (i.e., the line $y = x$), and about the situations where it would be "slightly harder." While Adam is no longer visually oriented exclusively to the upper edge, we can gather from his characterization of vertical counting as "another way" and his work on similar tasks that horizontal counting remains his primary method.

Episode 5 (Week 4, Task 4). In episode 4, we described how Adam recognized both vertical and horizontal counting as tactics for coordinating graphical displacements with changes in the intercept parameter of the equation $y = x + b$. We also suggested that Adam favored horizontal counting over vertical counting. In task 4 of week 4, we find his priorities changing. In this task Adam is presented with the grid view on paper and the lines $y = 2x + 7$ and $y = 2x + 1$ predrawn and labeled. He is directed to graph both equations with the computer and to "try to find the equation of a line which is half way between the lines you obtained." As Adam begins this task he recognizes that horizontal counting won't work "cuz it's a weird slant." He then draws in the line $y = 2x$, presumably as an analog to the reference line $y = x$ in previous tasks. In this episode, the horizontal displacement "doesn't look like seven" to Adam, and he confirms this by counting this displacement. But when Bluma says "that seven sort of pops up somewhere," Adam quickly orients to the vertical displacement and offers an account that inverts his previous assessment of the priority of horizontal over the vertical counting. Figure 5.4(b) shows the horizontal and vertical displacements between lines $y = 2x$ and $y = 2x + 7$.

From our perspective, this episode shows an important shift in the development of Adam's disciplined perception. From a mathematically disciplined perspective, vertical displacements are coordinated with the intercept parameter (b in $y = mx + b$), and Adam now appears to share this way of seeing with the grid view. In fact, the combined effect of the task with the "weird slant" and the tutor's act of disciplining ("that seven sort of pops up somewhere") leads Adam to describe vertical counting as "the reliable way." Despite this significant development in his mathematical perception, it remains to be seen what will happen if the grid view itself becomes unavailable.

Episode 6 (Week 4, Task 10). To this point, Adam's work has been consistently supported by his access to the grid view. Throughout the first 5 episodes, his uses of the grid view (and the visual aspects these uses relied upon) became more elaborate. Except in episode 3, when Bluma partially blocked his access to the grid's edges, the grid view has been available as a visual resource for these tasks. In the final 3 episodes, this vital resource is denied to Adam by the task environment. Just as in episode 3 when Adam announced that Bluma's blocking of grid view edges would result in "a different graph," removing the grid view produces a major epistemic disruption.

In this episode, the student is asked to make a prediction about the line $y = 2x + 3$ and then to graph the equation with the computer. No view, grid or otherwise, is shown on paper, but after he graphs the equation with the computer he can see the line and read the intercept in the grid view. The text relevant to this episode reads, "One can certainly read from the graph where the line crosses the x-axis. Is it possible to get that information from the equation (without the graph)?" Adam's first response to reading this text is to claim that it is possible by drawing a line. However, he then realizes that the task is directing him to find the intercept without the graph and he appears to be confused. Bluma points to the x-intercept of $y = 2x + 3$ on the computer grid view and asks him where it crosses (turn 1, below). To Adam it "looks like" negative one point five (turn 2, below), but he says he doesn't know how to do this without the graph.

1 Bluma: Uh-huh. So where does it cross?
2 Adam: Well, negative, looks like one point five. (1) "Is it possible to get that information from the equation without the graph?" What is it asking? To get that, to get what the x is?
1 (reading)
3 Bluma: Yeah, to get that negative one point five.
4 Adam: Maybe the x half the y. From where it crosses. It's probably not right. No. No-no-no. I don't know how.
5 Bluma: You don't know how? OK.

Immediately, Adam moves to disqualify the question itself, claiming that it "doesn't even matter if you could" "figure it out" without the graph (turns 6 and 8, below), because he can "figure it out" if he can "make the line" (turns 8 and 10). Bluma's reaction is astonishment, repeating his claim with emphasis on "DOESN'T MATTER" (turn 7). Bluma asks him to consider the possibility that it would be hard to find the intercept by making the line, and when Adam again says he "wouldn't know how," Bluma hints that "you know how to figure out this point" (turn 13). The episode closes as Adam proclaims in an exasperated tone, "I have no idea" (turn 18).

6 Adam: I don't even know if you can. Doesn't even matter if you could.
7 Bluma: It DOESN'T MATTER if you could?
8 Adam: You could just draw a little line (1) and then figure it out. You don't really need that.
1 (draws an imaginary line in the air)
9 Bluma: You don't really need to figure it out from the equation?
10 Adam: Yeah. If you can figure out where it crosses the y-axis, then you can make the line. You don't need to do it from the equation.
11 Bluma: Yeah, uh, let's see, what if it was sort of hard to make it from the line?
12 Adam: Well I wouldn't know how (sighs).

13 Bluma: OK, so how do you know, (2 sec) let's see, hmm, (2 sec), but you know how to, yeah, you know how to figure out this point (1) (3 sec), the y-intercept. (5 sec). Hmm.
1 (points to the y-intercept on the graph)
14 Adam: So, yeah?
15 Bluma: So, yeah?
16 Adam: What do you want me to do?
17 Bluma: So, but what, but (8 sec)
18 Adam: (exasperated) I have no idea.

We interpret this episode as a serious breakdown, not only in Adam's capacity to make progress on the task, but also in his interaction with the tutor. Turns 13 and 17 represent Bluma at her most inarticulate over the entire span of the tutoring sessions; she is clearly confused by his confusion. The task, by deleting the grid view, has brought two differing senses of "figuring it out" into direct conflict. Without the grid view, the intersubjective differences in Adam's and Bluma's meanings come into relief. In the final two episodes, we will see how these differing senses are mutually acknowledged and how Adam finally recognizes that there is a way "to figure it out" without the grid view.

Episode 7 (Week 5, Task 1). In this episode, Adam is presented with a blank view on paper, two parallel lines without names, and six points labeled A, B, . . . F (see Figure 5.1[b]). He is directed to answer questions about the two lines and to attempt to determine the coordinates of these points given only that the two lines graphed correspond to two of the following four equations: $y = 2x + 6$, $y = 2x - 2$, $y = -2x - 2$, and $y = -2x + 6$. As Adam reads the questions, he proposes that he can count (turn 1, below), but "since there's no lines on the thing it's really hard."[6] Since grid lines are absent, Adam begins to draw them (turn 3), asking Bluma, "Is that what they are telling me to do?"

1 Adam: "Find the coordinates of points A, B, C, and D." [Inaudible.] OK. The coordinate points, uh. Well, do they mean like through the equation? I can just go like this (1) but since there's no lines on the thing it's really hard.
1 (makes counting motion up and down with pen)
2 Bluma: Uh huh, yeah.
3 Adam: Is that what they're telling me to do, like go like this? (1)
1 (begins to mark points on x-axis between two lines and then gestures along the x-axis to the right)
4 Bluma: Um. Well there's (1) =
1 (begins to point to something on the page)

5 Adam: Is there a way to figure [it out?]
6 Bluma: [There's a] way to figure it out, right.

Adam's actions in this episode speak louder than his words. If only his critical resource, the grid view, were restored, Adam could count his way to a solution. He invokes an unnamed "they" wondering aloud if restoring the grid is what "they" are asking him to do. Adam does not follow through with this restoration and is unable to complete the task. But when he asks if there is a way to "figure it out" and receives affirmation from Bluma, we understand him to recognize that even without the grid view, the task can be done. Only later in the session, during our final episode (below) does Adam recognize and articulate a way to do this.

Episode 8 (Week 5, Task 3). In this final episode, the task is even more materially sparse and consists only of the following text: "In her notebook, Sue wrote the following: the point (3, 8) lies on the line $y = 2x+p$. Can you decide what number goes after the plus sign? Explain how you did it." Because Adam is confused by the variable p as he gets to work on this task, he and Bluma agree to replace it with 5, and then they consider the task in terms of a general process for determining whether the point is "on the line without the computer." As in episode 7, Adam again begins to restore the grid view, this time getting as far as drawing coordinate axes and tick marks on these axes (turn 4, below). As he begins to describe how he would use the drawing to "see," Bluma interrupts and in what we take to be another major epistemic shift, Adam realizes that there is indeed a way to figure it out without the grid view (turn 6, below).

1 Bluma: So how would you know to check that the three, eight is on that line without the computer?
2 Adam: The what, the three, eight?
3 Bluma: Sorry, yeah, three, eight. How do know to check that it's on this line?
4 Adam: Well, I'd draw, (inaudible) I'd draw two x plus five (1) one two three four five. I'd do that and see if three was on eight =
1 (draws two coordinate axes, marks points along the axes, and then draws in a line)
5 Bluma: = But you wouldn't know to =
6 Adam: = Oh, well, I could also go two times three plus five is eight. OH [emphatically]. That's what I should be doing. Ahh. Why didn't I do that? There we go.

In our view, this final episode marks an advance in Adam's development of a mathematically disciplined practice of using Cartesian space. He started with a robust and sometimes effective set of visual practices with the grid view that we have called the grid calculus (episode 1). These practices were elaborated, with the help of the tutor, to include an important visual aspect we have called "edges" (episode 2), and these edges were selectively effective for dealing with tasks in which horizontal and vertical displacements between lines were identical – lines with a slope of 1 (episode 4). This coordinated set of visual practices, contingent upon the presence of the grid view, was challenged in several ways: first by Bluma's attempt to block out edges of the grid view (episode 3), then by the introduction of tasks involving lines with slopes other than 1 (episode 5), and finally with the introduction of tasks that deleted the grid view and replaced it with what we call a blank view (episode 7). The final, decisive challenge came when Adam was denied visual access to Cartesian space altogether (episode 8). As the case closes, Adam suddenly realizes ("OH. That's what I should be doing. Ahh," at turn 6, above) that there is another way to "figure it out" without relying on the grid calculus.

In considering these tutoring interactions, one question we want to explore concerns the specificity of what Adam "knows" at the end of the sessions. To explore this question, we close our analysis of this case with a "thought experiment" which imagines taking Adam and his grid calculus to school to meet a traditional assessment.

Thought Experiment I: Taking Adam and His Grid Calculus to School

Let us begin by imaginatively transporting Adam to a middle school classroom on the day of an important test. Imagine further that in this classroom, the unlikely conditions of the tutoring experiment are replicated for each student: daily interaction with an expert mathematician around a well-conceived sequence of curricular tasks using a flexible computer-learning tool. Our attention is on a single student, Adam, as he prepares for the test. His confidence is high, because for three days he has become very good at using what we've called "the grid calculus." (We imagine the test coming at the end of episode 4, where Adam has provided an articulate account of the grid calculus.) The teacher circulates, handing out the tests, and as the test booklet is laid in front of Adam, a look of panic crosses his face. Quickly leafing through the test pages, Adam finds no grid views. In one question, there is a graph drawn on a blank view, but every other question directs him to use equations to determine intercepts and sketch lines with labeled points. Under these circumstances, it's not hard to imagine that Adam, who appears to us in these tutoring episodes as sharp, inventive, and "on task," would perform poorly on this important test.

A possible response to this imagined situation is to claim it would never happen; the resources provided by a test would never be that poorly matched to the tasks that preceded it. We invite the reader to imagine all the ways that such a mismatch could occur. For example, it seems unlikely that the author of this test, working with a coordinated mathematical perception of the Cartesian plane, would expect grid views provided in daily tasks "as visual aids" to become *the* necessary tool for a student instead of the equations. Thinking that the curriculum has been organized to teach the student how to use the equations and develop the "Cartesian connection," wouldn't the author be justified in believing he/she has written a reasonable test? And might not he/she be confused by the student's failure? Now imagine that every student in the class has worked as hard as Adam to develop a sensible and adequate approach to this particular sequence of mathematical tasks. Do we think that Adam is the only student for whom this particular test would be a design for failure?

The purpose of this thought experiment is to make a conceptual point about specificity and a political point about tests. Specificity is not a property of tasks or of practices but a relation between the two. In a world of tasks with slopes of one, the boy with the grid calculus gets along fine. Similarly, bringing the apparatus of "rational problem solving" to bear on shopping tasks in a supermarket is like cracking a peanut with a sledge hammer (Lave, 1988, p. 172). Scribner and Cole (1981) made a similar point about psychological tests to combat the widespread belief that tests could accurately represent the cognitive capacities of a culture. Our present concern is with the social implications of the specificity relations between mathematical tasks and assessments in the classroom, rather than ethnocentrism in cross-cultural characterizations of cognition. As a research community invested with responsibilities for creating assessment systems, we need better ways to understand specificity as a relation between tasks and assessment practices, since testing instruments function so critically to allocate social opportunities in this culture (Hanson, 1993).

Leaving Adam for the moment, we present a second case of disciplined perception, this one drawn from a workplace where Cartesian visual practices are also central to participation. In case 1, we showed how a student began learning to see in Cartesian space by interacting with a tutor and working through a specific set of school mathematics tasks. In case 2, we show how roadway engineers use Cartesian space to see with plan, section, and profile views. We present case 2 with two purposes in mind. One is to provide a contrasting case of disciplined perception. The other is to provide a backdrop for returning to the question of the specificity of what Adam "knows" at the end of his tutoring sessions.

Case 2: Using Cartesian Space to See

Our two engineers, Jake and Evan, are differently experienced with civil engineering, but not nearly so differently experienced as were the tutor and student with Cartesian coordinate mathematics. So, while we organized the first case to demonstrate how the student's perception developed and changed, we build the second case around the assumption that we are documenting a relatively stable form of disciplined perception in roadway engineering. And, again, what we mean by this is the engineers have developed a visual practice in which aspects across views can be coordinated to accomplish the tasks at hand. In this case we (1) demonstrate this coordination over the course of a single naturally occurring problem-solving event, (2) argue that the two engineers more or less share a disciplined perception that is reflected in their interaction during this event, but also (3) analyze an important difference between how the two engineers orient to a specific design aspect, a difference that leads the more senior engineer to discipline the perception of his junior colleague.

Background

This case documents two professional civil engineers working to redesign a roadway for a proposed housing development in northern California. We recorded four continuous hours of videotape at their firm, mostly in the personal workspace of one of the engineers. During breaks in their activity, we informally interviewed the engineers about their work and about this project. We also interviewed them more extensively before and after making this video record of their work. Jake is the current project manager on the housing development, and Evan is a more junior engineer working for Jake on this project. They are revisiting this project after a 6-month hiatus; their charge on the afternoon we visited is to make revisions to a roadway design plan after receiving some new information about the site and to recalculate some quantities with a computer-aided design (CAD) system to reflect these changes.

Jake and Evan are primarily concerned with quantities they call "cut," "fill," and "slope." "Cut" and "fill" are measured in cubic yards and refer to the amount of earth to be removed or added at a particular location in a development plan. Unlike in school mathematics, "slope" is represented in percentages. Maximum allowable roadway slopes in the project are either 15 or 20%, depending on the kind of roadway in question. There are many general constraints that bear on this particular design. Besides meeting codes that specify physical features like maximum allowable slopes, the engineers try to "grade" a site (i.e., reshape the terrain to bring "existing" into alignment with "proposed") in a way that "balances" cut and fill (i.e., the net sum of cut and fill across the site equals zero). Balancing the site is important primarily for economic reasons; both "exporting" and "importing" dirt are costly activities. The roadways also must be designed with the entirety of the development project in mind. For example, in order to begin building homes, roadways must provide access to the locations where the houses will be built.

There are other important constraints on this particular project. For example, the building site is especially hilly and this makes grading difficult. As an example, one major concern is the placement of roadways along a steep slope. In a situation like this, cutting into a hillside presents problems, since cutting too much at the bottom of a slope leaves the remaining slope inadequately supported. The common solutions are either to "chase grade" by cutting off the whole hillside, which can have significant environmental consequences, or to put in an expensive wall to support the hillside. In this project, the engineers are required to jointly manage the general constraint of "balancing the site" and the specific constraint of producing a design that is environmentally sensitive. "Minimizing grading" in light of these environmental considerations "drives" this design for Jake and Evan.

Civil roadway engineering involves the coordinated, embodied use of three primary views: plans, profiles, and sections. At this firm, these views are produced with a CAD system. Before the advent of such systems, these views were drawn by hand, and in an interview, Jake recounted once having to draw sections at a site in harsh Massachusetts weather. Now they use the computer to "do a roadway," starting with a plan view to "get a sense of how the land lies" and to "snake" a proposed roadway through the existing terrain.[7] Given the plan view, they next create a profile or elevation view that allows them to "snake the vertical alignment through" the development site. Finally, "to see how it worked," they make section views that show vertical slices through the roadway at intervals of 50 feet (see Figure 5.5).

Readers of this paper should expect to have difficulty in following the visual practices of civil engineering, even though these practices make

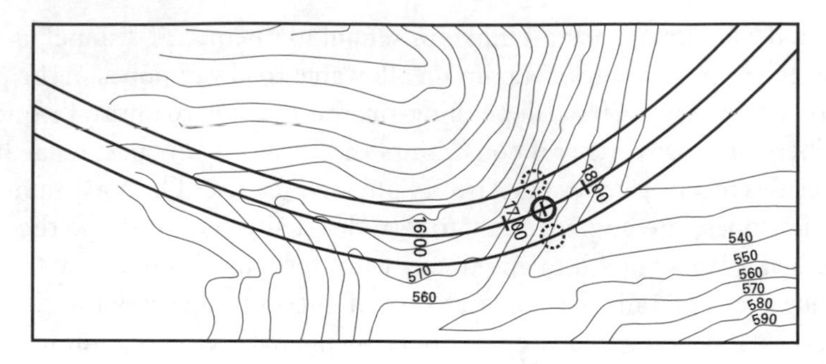

5.5(a) Plan view with stations and topo

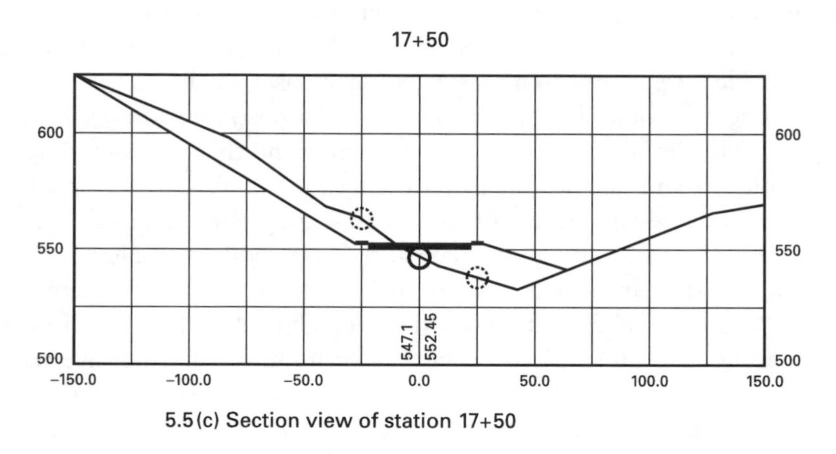

5.5(c) Section view of station 17+50

Figure 5.5. Three views of a roadway: plan (a), profile (b), and section (c).

use of conventional Cartesian coordinate systems. Accordingly, we carefully explain the specialized representational views of civil engineering before proceeding with our analysis. We invite the reader to consider that the initial strangeness and difficulty this system of representations may present roughly parallels what the student Adam is faced with in making sense of the views he encounters.

In Figure 5.5 we have selected portions of plan (Figure 5.5[a]), profile (Figure 5.5[b]), and section (Figure 5.5[c]) views to show the reader how these engineers coordinate their visual orientations within and across these views. The first thing to notice is that the views share a common quantitative scale that allows a user to find the same location across views. These quantities, represented using slightly different notations across the views (e.g., "17 00" on the plan versus "17 + 00" on the section and

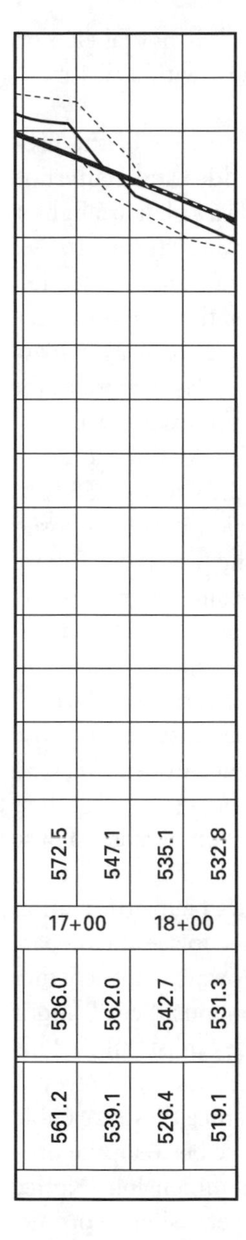

5.5(b) Portion of profile view around station 17+50

profile), are called "station numbers." They are generated by picking an arbitrary starting point (usually the beginning of a road) and then placing a station every 50 feet along the road. Station "seventeen fifty" in Figure 5.5 is $17 \times 100 + 50 = 1{,}750$ feet from the beginning ("station zero") of this particular road. Notice also that while "17 + 50" is not labeled on the plan and profile views, it is indicated, respectively, with a tick mark and a ruled vertical line.

The plan view (Figure 5.5[a]) provides a bird's-eye view of the proposed roadway. The three dark lines represent the sides and centerline of the proposed roadway. Layered beneath the roadway plan are lighter topographic lines that indicate planes of constant elevation (every 10 feet above sea level) for the existing terrain. By orienting to the intersection points of the roadway lines and the topographic lines, the reader can see the existing elevation of the terrain where the proposed roadway will go. We've circled these three points in Figure 5.5(a) using dashed circles for the sides of the road and a solid circle for the center. For example, by following the "550" topo line, the reader can see that the existing elevation at the proposed centerline of station "17 + 50" is slightly below 550 feet.

The profile view (Figure 5.5[b]) represents a side view of the proposed road. (Only a sliver of the entire profile is shown in the figure; see Figure 5.6[a] to see more of the whole profile.) The thicker solid line represents the proposed roadway's elevation; the thin dashed lines indicate the elevation of the existing terrain at the two sides of the proposed road; and the thin solid line indicates the elevation of the existing terrain at the centerline of the proposed road. The entire profile is set against a grid whose vertical axis is labeled every 100 feet and grid lines are drawn every 50 feet. The horizontal axis is labeled with station numbers (e.g., 17 + 00), and three additional quantities appear near the horizontal axis. These represent the existing elevation at the centerline (e.g., 572.5 above the station 17 + 00) and at the sides (e.g., elevation values below the station 17 + 00). Notice that the profile view allows the user to see the disparity between the existing terrain and the proposed roadway. If the existing terrain is above the proposed roadway, the developers must "cut," and if the existing terrain is below the proposed roadway, they must "fill." This disparity is not visible in the plan view. The profile does not, however, allow the user to easily see the amount of cut and fill since these are cubic quantities, and the profile view only shows a difference between the proposed roadway and the existing terrain in the vertical dimension. Notice also that the curvature of the proposed road is no longer visible in profile, as it was in the plan view.

The section view (Figure 5.5[c]) represents, in Jake's words, "a slice through the road" every 50 feet. The thick horizontal line represents the proposed road surface and the thinner solid lines represent side views of both the existing terrain and proposed roadway at this station. Again, if

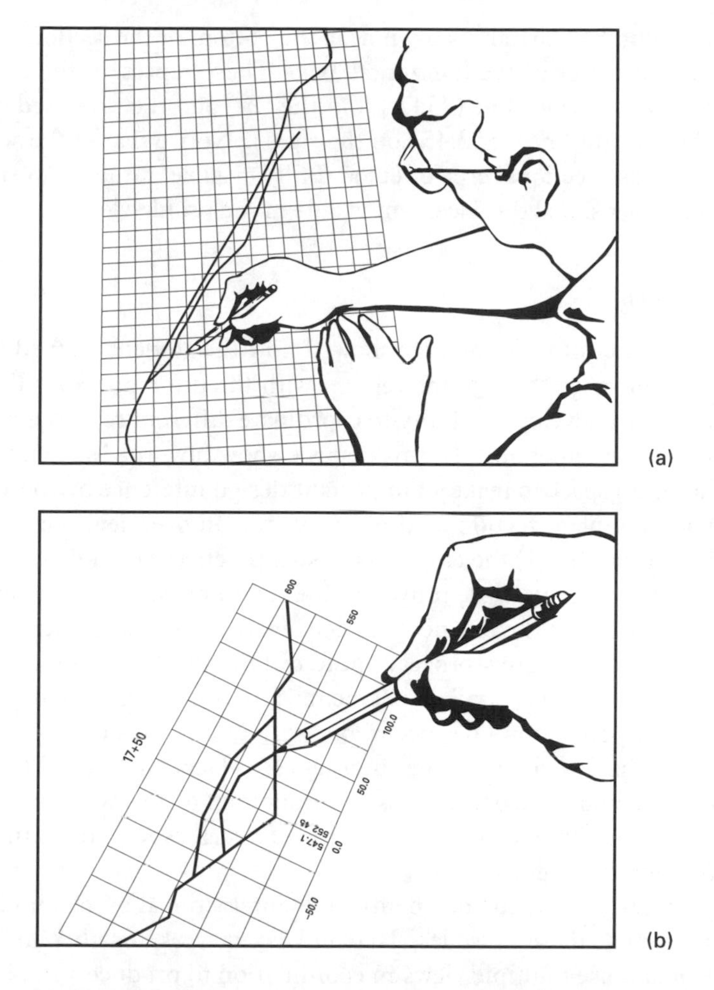

Figure 5.6. Jake makes different aspects of an alternative roadway visible for Evan across two views. In Figure 5.6(a) Jake traces the trajectory of the alternative roadway on the profile (turn 27, action 1). In Figure 5.6(b) Jake shows Evan how this alternative would appear in section view (turn 38, action 2).

the existing terrain is above the proposed roadway then they must cut, and if the proposed roadway is above the existing terrain then they must fill. Section "17 + 50" represents a station where the developer must both cut (i.e., region to the left) and fill (i.e., region to the right). Sections are also set against a grid that establishes a horizontal axis with positive and negative distances out from the proposed road's centerline and a vertical axis, again with elevations in feet above sea level. Two elevation numbers are also shown near the bottom of the section on two sides of the center of the horizontal axis. These represent the existing elevation at the centerlinc ("547.1" on the left) and the proposed elevation of the centerline ("552.45" on the right). Notice that in the section view the cross-sectional area of cut and fill is visible, rather than simply the one-dimensional displacement visible in the profile view.

Analysis

In the tutoring case, we showed how the student learned to see in Cartesian space through interaction with the tutor, an expanding set of tasks, and a developing ability to coordinate different representational views of linear functions. In this case we show how roadway engineers use Cartesian space to make an important design inference by coordinating aspects of plan, section, and profile views. In overview, our analysis of this case shows: (1) the engineers' visual practices are, unlike the student's in the tutoring case, fully coordinated to pursue the relevant engineering tasks; (2) the engineers largely share a disciplined perception, and so their interactions proceed quite differently than those in case 1 (i.e., few breakdowns, similar judgments and assessments); and (3) disciplining of perception does occur, but with less serious epistemic consequences than we found in the tutoring case. Here we find a difference in judgment about a particular visual orientation to a view, rather than a different commitment to the very status of visual aspects (e.g., the grid points along the edge) and their use.

In this case, we divide one contiguous interactional event into 5 episodes. Across these episodes, Jake and Evan work together to find a problem and use multiple views in coordination to produce a resolution. Episodes in case 1 were used to show how disciplined perception changes over time and through interaction between a tutor and a student. In contrast, episodes in case 2 are used to show how disciplined perception is deployed to handle a naturally bounded task. The problem Jake and Evan find and resolve involves the familiar Cartesian topic of slope. However, even within this task, we will find that breakdowns do occur.

Episode 1: Finding a Problem. Jake and Evan are revisiting the roadway design after 6 months. As they begin the work of changing roadway designs and recalculating quantities, they discuss the slope of Road 4. Evan remembers that they "went with a max slope" (15%) and Jake remembers going with "more than fifteen percent."

1 Jake: Yup (1). Yeah, That's all we have for Road Four. (2 sec) Yeah, and I kinda remember that, (2) I remember it was after the fact and they just wanted us to look at it and (3) we basically just ran the road up the hill to the extent that we cou =
1 (Jake crosses the room)
2 (sharpens a pencil)
3 (returns to table)

2 Evan: = I remember you were, yeh, you pretty much left it up to me. You said, you know, just (1) you know, get some (2), we went with a max slope HERE (3), and that's what I did, then I put it, what one of those (4) =
1 (flat R hand moves over surface of profile and retracts)
2 (R finger points to profile)
3 (lays edge of R hand flat along road in profile)
4 (R finger moving back and forth along segment of proposed road)

3 Jake: = Actually, I think this (1) is even. . . . Now thinking back I think we went with MORE than fifteen percent (1 sec)
1 (holds L finger on road in profile)

4 Evan: On this one?

5 Jake: Yeh. Yeh, hah . . . cuz do you (1) remember didn't we find that fifteen percent just wasn't gonna (2)
1 (walks across room and gets calculator from desk)
2 (begins using calculator)

6 Evan: Yeah, yeah, we may have. I remember, I think I remember being really concerned about it.

Because the "city design guidelines" allow a maximum slope of 15% on this kind of road, Jake and Evan have a potential problem. Since they cannot read the slope directly from the profile view, Jake gets a calculator to "look" at the slope, using a version of the conventional slope formula (delta y/delta x) to find that the road's slope is close to 20%.

7 Jake: Let me just look. (1) (inaudible) forty two point one minus, five (2) Yeah, we went twenty percent. (2 sec) Member? (3)
1 (L finger points to existing profile view and R hand keys numbers into the calculator)
2 (hits enter key on calculator to complete the computation)
3 (writes 20% on profile near road)

8 Evan: Yeah, I remember. (3 sec) (1) I re . . . ah remember not thinking that they were too serious about it, at that time.
1 (Jake begins another calculation)

9 Jake: (1) (2) (3) (4) Umm, nineteen and a half, whatever. (3 sec) Well that's alright, we can just (5) note that. (sighs) HAhh.
1 (using calculator)

2 (circles existing elevation – "606.2" at station "14 + 00" in profile)
3 (circles existing elevation – "353.1" at station "27 + 00" in profile)
4 (using calculator)
5 (takes glasses off, rubbing eyes)

10 Evan: That's the city criteria?

11 Jake: Yeah . . . city design guidelines. It's fifteen percent max on, on uh . . . collector roadways. (1) Secondaries they allow you to go steeper. I think they actually allow you to go to twenty.
1 (looks at Evan)

12 Evan: Oh OK. Yeah, well.

13 Jake: (1) But that's alright. (2)
1 (Jake looking at Evan, 5 seconds)
2 (looks back at profile)

Episode 2: Looking for a Design Rationale. Although Jake says the situation is "alright," they face the problem of explaining why Road 4 might be acceptable with the excessive slope. Seeking a rationale for the excessive slope, Jake leads a coordinated tour of the profile, plan, and section views of the proposed road. By orienting to station and elevation quantities on the horizontal and vertical axes, Jake realizes that this road is going to be "brutal," rising 20% over 1500 feet. In his first bid to explain the excessive slope, Jake orients to a roadway segment in the upper part of the profile view and then shifts quickly to the plan view with topographic lines, recognizing that "the slopes are so steep" at this upper end of the canyon. This leads him to suggest that they "wanted to avoid getting into a lot of cut" at that end of the canyon since steep slopes mean the likelihood of "chasing grade." However, this reasoning "doesn't seem right" when he looks at this segment of road within a field of topo lines and finds the road to be on one side of "the low point." Throughout this episode, Jake produces coordinated readings across the views by using the common index of station numbers.

Since Jake can't find a plan view that includes stations and topo in his stack, he has to coordinate aspects across two different plan views – one with roadway and topo (no stations) and the other with roadway and stations (no topo). He uses his hands to coordinate across the three views. Doing this allows him to examine the topographic conditions at specific stations, which is particularly important in his search to find a design rationale for a "brutal" slope.

We interpret the transition from turn 21 to turn 23 as an artful act of disciplined perception. In action 2 of turn 21 Jake uses the profile view to orient to a span of stations around "16 + 50," which is a location where they will need to cut. Jake carries this span of stations into the plan-with-stations (action 3) and then into the plan-with-topo (action 4) to find a region they will have to cut "anyways." Since the station numbers are not present in the plan-with-topo, Jake must orient to some visual aspect of the road in the plan-with-stations (e.g., its shape or its curvature). He carries this aspect into the plan-with-topo, in order to mark out "this one area" (turn 23) where they will cut. This entire coordinated assembly of aspects across representational views (i.e., profile, plan-with-station, plan-with-topo, and Jake's pencil drawing) occurs within just two turns at talk.

20 Evan: You want to know about (1) this station?
  1 (points to a station on plan-with-stations)
21 Jake: Yeah. See like right here (1) (5 sec.) Like right at sixteen (2) fifty or so (3) like right at this corner (4), we're (5) actually gonna cut, we're gonna have to, we're gonna have to . . . take down (6) the side of this anyways, you see that?
  1 (overlaps and layers views with profile on top, plan and stations next, then plan-with-topo)
  2 (L finger points below "16 + 50" in profile)
  3 (L finger moves to "16 + 50" on plan-with-stations and holds)
  4 (R pencil tip moves to corresponding spot on plan-with-topo)
  5 (L hand moves back to profile at "16 + 50")
  6 (scribbles over region above road in plan-with-topo)
22 Evan: Yeah.
23 Jake: But, um . . . That will be like . . . it's just in this one area. (1) So it'll be somethin' like, you know, this. But then see here (2), I think what was goin' on is, I think here. This is all (3) trees, right?
  1 (R pencil draws dashed line in area next to road on plan-with-topo between stations "14 + 50" and "18 + 50")
  2 (L finger moves back to road on profile around station "14 + 50")
  3 (R pencil traces a circle around the area surrounding the road in plan-with-topo between stations "10 + 00" and "14 + 00")
24 Evan: Um hm.

Episode 3: Finding a Design Rationale. Finding the region around "16 + 50" in the plan-with-topo view allows Jake to move along Road 4 to consider a region to the left that is "all trees." Here his argument about avoiding cut will be used again, but this time with more success, producing the beginning of a "reason" for why they designed the road with the excessive slope. "Even five feet of cut" in this region will require that they "chase grade all the way up on both sides." As Evan (and we) learn at the end of this problem, this would violate one of the project's main directives: avoid denuding the tree-covered hillsides for environmental reasons. As Jake narrates this reason, he animates each element for Evan: the region that is all trees (R pencil traces a circle around the area surrounding the road in profile between stations "10 + 00" and "14 + 00"); that they could have to chase grade on both sides (pencil moves up and away from road in plan-with-topo near station "13 + 00," then pencil moves down and away from road in plan-with-topo near station "13+00"); and the regions they will have to grade (pencil repeats the previous two movements with added sweeps across the wider hillside).

Episode 4: Strengthening the Rationale. Jake now has a reason they didn't "flatten [the road] out," framed in contrast to an alternative design that he and Evan collaboratively narrate and animate. Jake still needs to do more work to strengthen the validity of the road's current design. This road may be objectionable not only because of its slope but also because it will require a substantial amount of fill in certain regions. Jake works to mitigate this potential concern by moving back to the profile view and orienting to the stations where fill is greatest (around "23 + 50" and "24 + 00"). Jake's hands come quickly up off the surface of the two-dimensional view to animate "the canyon's comin' up HARD" (arms come up abruptly to gesture a wide "v" above the profile), vividly displaying the fact that the "actual area of grading is gonna be . . . not that significant." When Jake asks Evan if "you know what I'm saying?" Evan quickly and repeatedly agrees. When Jake draws a section that illustrates this situation in the canyon (draws a section on a blank region of plan-with-stations), Evan calls the section a "quick fill" and then elaborates, showing that he is also oriented to the relevant aspects this illustrative section is designed to show (i.e., "deep but not encompassing a lot of area").

Episode 5: Jake Disciplines Evan's Perception to Keep the Rationale Strong. While Jake and Evan appear to agree that it's only going to be a "quick fill" in even the worst sections, they turn to the actual sections nonetheless, presumably to check their claims. With the sections visible, Jake orients to the horizontal displacement of the section "23 + 50," instructing Evan to "look at" what he apparently considers a confirmation of their claims that the "actual area" for the deep fill isn't that significant ("only a hundred and ="). When Evan actually sees the section, however, his initial assessment is different. Echoing Jake's earlier use of the evaluative term "brutal," Evan points to the area between the proposed and existing in this section view, tentatively stating, "That's brutal, huh?" This assessment leads Jake to discipline Evan's perception, telling him to "look at it this way" as he sketches alternative versions of the roadway sections "23 + 50" and "24 + 00" (see Figure 5.6[b]). These sketches allow him to contrast the gain of "a little bit less fill" against the loss of "ripping down the whole side of the hillside." Having transformed the problem of an excessive road slope into an articulate and defensible design rationale, Jake settles back in his chair and states "so that's why we did this the way we did it as I recall" (turn 42).

36 Jake: Right. See look at that (1) That's only (2 sec) a hundred and =
1 (points to numbers on horizontal axis of section "23 + 50," below the left and right sides of the roadway)
37 Evan: = (laughs) Boy, that's (1) brutal, huh? Well, uhh, maybe not.
1 (points to interior region of section "24 + 00")
38 Jake: = Yeh. Well, but, look at it this way. You can have THIS (1) or you can have THIS. (2) Big deal, right? Little bit less fill down here (3), and what that buys you is [at the]
1 (points to interior region of section "24 + 00")
2 (draws another small road section on "24 + 00")
3 (draws another small road section on "23 + 50")
39 Evan: [cut]
40 Jake: other end (1) you're ripping down the whole side of the hillside.
1 (with pencil gestures up and to the left)
41 Evan: Right.
42 Jake: So that's why we did this the way we did it as I recall. (1)
1 (settles back in chair, interlocking hands behind his head)

Comparative Discussion

1. Is Perception Socially Constructed?

With these cases, we argue that technoscientific perception is socially constructed. By this we mean the way technoscientists orient to and coordinate visual aspects is fundamentally shaped by their activities with other people and culturally specific artifacts. This is a conceptual point that historians and philosophers have made for some time (Hanson, 1969; Kuhn, 1970; Wartofsky, 1979), but following contemporary analysts (Goodwin, 1995; Goodwin & Goodwin, 1996; Goodwin, 1994; Latour, 1990; Lynch, 1990), we've tried to show this process in real-world technoscientific activity. We have tried to go further, particularly in case 1 but also in case 2, to document how technoscientific perception changes through interaction (i.e., how one kind of learning happens).

> To remark of some phenomenon or its properties that it has been "socially constructed" is not thereby to say that it has been relegated to the status of "mere" artifact or the "arbitrary" upshot of social consensus. The pejorative connotation of the concept of "artifact" in the experimental sciences and the fallacy of the "consensus theory of truth" in epistemology should not blind us to the fact that, in a significant sense, "objective findings," "absolute mathematical truths" or "reality-determinations" are rendered possible only by the acculturated, concept-laden and fundamentally social (because communicative and intersubjective) operations of their producers and consumers. The interest of this for epistemically-minded social scientists lies precisely in elucidating how this is so. (Coulter, 1989, pp. 19–20)

In case 1, "figuring it out" and "checking" are examples of the kinds of operations Coulter describes. Around these terms, mathematical intersubjectivity is being worked out. In case 2, Jake and Evan share a great deal more than Bluma and Adam do in their ways of using words and views, but even inside a civil engineering firm there are still relevant differences to be worked out, as when Jake challenges Evan's judgment of a section fill as "brutal." We see from case 1 that different senses of these terms can coexist in an ongoing conversation, coming into contention and alignment only when conditions are right. As a "tutoring experiment," the conversations in case 1 are framed to be explicitly pedagogical, but even so, participants in these conversations cannot ask at every turn whether they are using "figuring it out" to mean the same thing. What then are the conditions that allow contention and alignment?

Case 1 shows us that breakdowns in activity, when recognized, are such occasions. Breakdowns occur when routine ways of working are disrupted. And disruptions can appear at different places in a system of activity: (1) in the recurrent pattern of tasks (e.g., when Adam is asked to solve problems where his grid calculus won't work), (2) in the tools (e.g., when the grid view is deleted, late in the sessions), and (3) in the person (e.g., when Adam's visual access to the grid is blocked or he is directed to look at different aspects of the views). In case 2, the tasks and tools are relatively stable, but disruptions can still occur between people, more specifically between their judgments, as when Jake challenges Evan's assessment of the section fill as "brutal."

2. Is Conceptual Change Asymmetric in Interaction Between People?

In case 1, we focused on how the student's perception changes over the course of the tutoring sessions. Obviously learning in interaction is rarely (if ever) a one-sided phenomenon; when learning occurs, meaning is mutually appropriated. In case 1, for example, the tutor clearly comes to learn a good deal about an unfamiliar way of seeing the Cartesian plane. So in order to avoid giving an asymmetric analysis of learning must we say that the student also has disciplined the tutor? Doing so would risk allowing the concept of "disciplining" to mean too much and lose its usefulness. "Disciplining perception" refers to everyday interactional events that *are* asymmetric: those where one interactant works to reorganize another interactant's way of visually orienting to the world. Thus, while the student defends his ways of seeing to the tutor, he does not issue directives to her. On the other hand, the tutor does issue directives and blocks access to views in order to provide the student with "another way to look at it."

From both cases, we can construct a limited but grounded theoretical model of "disciplining" perception as it occurs between people: (1) one or more interactants recognize an intersubjective disparity, (2) an announcement is made concerning the disparity, followed by (3) a directive of the type "look at it this way," which initiates (4) an embodied course of action through which the initiating interactant coordinates aspects of views to animate "this way." Often, but not always, (5) this alternative way of seeing is accompanied by an account which justifies it as, among other things, easier, less expensive, more useful, more precise, et cetera. Finally, the recipient of this performed course of action may (6) demonstrate a similar performance that rehearses the coordination of aspects in the alternative way and shows that he or she has understood the intended import of the speaker's actions.

3. *How Do Accountabilities Shape Disciplined Perception?*

The engineers in case 2 have different and often conflicting accountabilities (Strauss, 1985).[8] These include accountabilities to: (1) cost-effective design practices, (2) the city, and (3) the client. These different accountabilities can be traced to different forms of visual coordination, respectively: (1) coordinations to see cut and fill, (2) coordinations to see whether roadway slopes meet or exceed code, (3) coordinations to see whether they will chase grade in a tree-filled canyon. This linkage between accountabilities and the way they are actively perceived in the materials can help us understand the act of disciplining perception in case 2. When Evan announces the fill at section "23 + 50" is "brutal," he is demonstrating a coordination that addresses only one of these accountabilities. When Jake disciplines Evan's perception, he performs two different coordinations together: demonstrating the minimal gain ("no big deal") of meeting this single accountability against the potential failure to meet a more important accountability to the client (not "ripping down the whole side of the hillside"). In case 1, it appears that the primary accountability that shapes Adam's developing disciplined perception occurs face-to-face in interaction. It is Bluma who actively represents to Adam certain accountabilities of disciplined mathematical practice, such as exactness (episodes 1 and 2) and "figuring it out" or "checking" with the equations (episodes 1, 6, and 7).

Finally, to close our discussion, we propose a second thought experiment symmetric to the one that closed the first case analysis. Again, our point in conducting these thought experiments is to raise questions about the specificity of Adam's knowledge by asking how well or poorly this knowledge is aligned with particular settings. In this case, we imagine Adam a little further along in his development of schooled Cartesian practices and invite him into the workspace of our two engineers.

Thought Experiment II: Taking Adam and His Equations to Work

This thought experiment imaginatively transports Adam and his late developing use of equations to the civil engineering offices of Jake and Evan. As case 1 ended, we showed Adam moving away from using the grid calculus and toward the use of equations. Let us suppose that this developmental trajectory continues in his middle school math classes. If we take a standard algebra textbook[9] as a measure of the kinds of tasks Adam will encounter in these classes, an equation-centered set of practices will serve him much better at school. We imagine Adam becoming very good at using equations to meet task demands in the classroom and that Adam gradually forgets the grid calculus since he no longer uses it.

Adam is now sitting with Jake and Evan, inspecting their excessively steep roadway. Our question is: Will Adam be more capable in this setting with an equation-centered practice than he would have been if he had continued to develop his grid calculus? We propose that Adam's uses of the grid view might put him in a better position in this setting than an equation-centered practice. In many ways his coordinated uses of the grid view are quite similar to what Jake and Evan do with their views, though much simpler. In both cases, visual aspects are *read from* inscribed

Cartesian spaces and coordinated with quantities. Jake and Evan carry quantities across multiple views and make complex inferences; Adam uses a single view and does simple, single-answer tasks. Nevertheless, it's the similarity of use that we want to highlight. In both cases, it is the views, and coordinated readings from them, that are trustworthy resources for making inferences. In both cases the computer is used to produce visual forms that can be used to make subsequent inferences. In neither case are calculations from equations the primary practical resource.

It is this quality of *reading from* the views that we believe distinguishes these ways of using Cartesian space from uses in the discipline of mathematics. For Bluma, and other mathematicians, we believe that tasks *within the discipline of mathematics* lead the user to treat equations as the trustworthy resource *in this kind of mathematical situation,* since the use of equations authorizes inferences about points and lines in Cartesian space.[10] This is why, in case 1, Bluma as a mathematician uses the term *checking* to refer to an equation-based practice (e.g., asking Adam, "But did you check it in your head?"). In contrast, Adam uses *checking* to refer to a view-based practice. This is also why Bluma tries to block Adam's visual access to the grid view (episode 3), to challenge his use of the grid calculus.

To those who are mathematically enculturated, Adam's statement that the blocked grid view is a "different graph" is surprising and possibly amusing, but this assessment of Adam depends on a particular interpretation of what he means by these words. If we interpret Adam's statement as being about ontology (i.e., he sees a different graph), then his difficulty is analogous to Evan's with a proposed fill. In episode 4, where Jake holds together elevation and plan views, Evan sees the proposal as a "quick fill" that is "deep but not encompassing a lot of area." However, in episode 5, where Evan sees the area of fill in isolation (i.e., in a section view, alone), this same fill becomes "brutal." If, however, we interpret Adam's statement as being about usability (i.e., the blocked grid view would be difficult to use), then the blocked view is a different graph in the sense that it cannot be put to the same uses as an unblocked one. Analogously, two plan views (one with topo, the other with stations) are different for Jake and Evan (as in episode 2) because they cannot be used to answer the same questions.

Conclusion

We started this paper with questions about (1) how people in school and work settings use coordinate systems of representation to reach shared understandings, and (2) how breakdowns in and between their talk and their visual practices lead to significant teaching and learning events. We now summarize our findings and point forward to how the kinds of comparative work reported in this paper can build productive connections between educational research and sociological/anthropological studies of technoscientific practices.

In our analysis of a tutoring situation (case 1), we found repeated breakdowns and several striking instances of the tutor working to discipline the perception of her student (e.g., directing his attention to "evenly spaced" lines in episode 2 and blocking his access to the grid view in episode 3). Under these circumstances, we also found ample evidence of development in the visual practices of the student. These were not uniformly in the direction of a normatively organized disciplinary perspective on graphs in the Cartesian plane. Rather, we found that the student's visual practices around what we called a "grid calculus" were both deeply entrenched (e.g., his preference for counting up/over along lines formed by grid points) and flexibly extended in response to the tutor's statements (e.g., his elaborated use of what we called "edges" at quadrant boundaries). The visual aspects used in this grid calculus were relevant to understanding how a coordinate system works (e.g., equivalencies over horizontal and vertical displacements), but they were also overly specific (e.g., counting horizontal displacements for functions with slopes other than 1), effectively trapping the student inside a set of visual practices that had limited scope or generality. Following systematic efforts by the tutor and curricular materials, the student did recognize a connection between his elaborated grid calculus and more conventional uses of equations (e.g., "OH. That's what I should be doing" in episode 8).

In our analysis of a design problem in civil engineering (case 2), we found a much more fluent and coordinated set of shared visual practices operating over a collection of conventional Cartesian displays (i.e., plan, profile, and section views). By assembling these views into coordinated representations of the difference between existing and proposed space, these engineers were able to find unresolved design problems (e.g., a "brutal" slope for a proposed roadway), to construct alternative designs (e.g., a less steep roadway through the same region of the development site), and to evaluate qualitative and quantitative differences between these alternatives (e.g., trading the cost of fill dirt against the environmental consequences of denuding slopes in the development site). By and large, their work proceeded without significant lapses in shared understanding.

However, when shared understandings did break down (e.g., the junior engineer evaluated the amount of fill below their proposed roadway as "brutal" in episode 5), we found processes of disciplining the perception of relative newcomers that were very similar to those in the tutoring situation. In both cases, old-timers made concerted efforts to reorganize the visual orientations of newcomers and new forms of coordination among visual aspects of Cartesian space were achieved.

Our analyses should not lead us to understand mathematically disciplined Cartesian practices and other disciplined Cartesian practices as incommensurable; it should lead us to see them as importantly different. Mathematical practices in general – the uses of space, quantity, and pattern – are richly varied across real-world settings. This variation invites attention from scholars of mathematics education. Reformist agendas across our research community will vary; some of us will seek to reorganize mathematics education to make it more relevant to students' everyday experiences, others will attempt to help students learn in ways that will enhance their future work prospects in disciplines like engineering, and still others will seek to develop classroom practices that allow students to participate in public institutions whose discourses are frequently technical and quantitative. Whichever agenda is emphasized, we believe a program of research that investigates the relations of specificity between embodied practices and tasks across diverse mathematical settings is a promising direction.

We take this chapter to be an early move in a larger collective project that seeks to bring together studies of technoscientific practice with research on teaching and learning in schools (Hall, 1995; Hall & Stevens, 1995; Hall & Stevens, 1996; Lehrer, Schauble, Carpenter, & Penner, 1996; Roth, 1996). Our analyses of "disciplined perception" represent a bridge between two communities of research that share an interest in representational practices but differ significantly in their foci. With a few exceptions, teaching and learning are entirely absent as thematic concerns in even the best research on professional technoscientific practice. On the other hand, educational research, with its recent cognitivist leanings, is only beginning to explore the diverse webs of interactions that comprise the lived work of technoscientific learning and practice. Our continuing research pursues a synthetic approach that displaces the great divide between technoscientific schooling and professional practice with a series of grounded studies that demonstrate how people manage the manifold transitions between these arenas.

Acknowledgments

We would like to thank Alan Schoenfeld and Andy diSessa for multiple types of support to Reed Stevens in developing earlier versions of this paper. Thanks also to the Functions Group at the University of California for sharing data for case 1 and to the civil engineers for sharing their time and space. We would also like to thank Cathy Kessel and Susan Newman for sensitive and helpful readings of the current version. Anne Schwartzburg's graphic expertise was essential to assembling the figures in the paper. Special thanks to Magdalene Lampert for editorial assistance. This work has also been supported by a NSF Spatial Cognition traineeship (UC Berkeley) to Reed Stevens and by NSF Grant ESI 94552771.

Notes

1. Following Latour (1987, pp. 174–175), we use the term *technoscience* – rather than *science, technology,* or *mathematics* – to reopen questions about what elements comprise these disciplinary practices as they are found in scholastic and professional settings.
2. We use the terms *inscription* and *visual display* as synonyms in this paper.
3. An extended discussion of "seeing aspects" and this activity's relation to language use can be found in Wittgenstein's *Philosophical Investigations* (1953). The term and the spirit of its use are borrowed from these writings.
4. See Schegloff (1984) and McNeill (1992) for two notable discussions of the relations between talk and the more restricted action category of gesture.
5. Figure 5.3 is a redrawn version of a "snapshot" from the original video recording (whose resolution was not high enough to enable reproduction here); likewise for Figures 5.6.
6. By "lines" here, we assume he means rows or columns of grid dots in the grid view.
7. A full visual and textual representation of Jake's embodied performance of roadway design can be found in an interdisciplinary artist's book entitled *Making Space: A Collaborative Investigation of Embodied Engineering Knowledge* (Schwartzburg & Stevens, 1996).
8. A distinctive sense of *accountability* can be found in the corpus of ethnomethodological writings – "when I speak of accountable my interests are directed to such matters as the following. I mean observable-and-reportable, i.e. situated practices of looking-and-telling" (Garfinkel, 1967, p. 1). While we use the term here in a more commonplace way – meaning something like recurrent institutional responsibility – Garfinkel's quote should make it clear that the ethnomethodological sense of accountability has shaped this chapter throughout. For elaborations on differing senses of accountability, see Newman (in press) and Stevens (in press).
9. For example, in the *Holt Algebra I* (Nichols et al., 1992), none of the coordinate geometry problems, except naming points on a grid, can be solved using the grid calculus. Grid views appear infrequently as visual aids. This trend appears to continue in textbooks that arrive later in the traditional mathematics curricular sequence.
10. As Cathy Kessel has suggested to us, mathematicians coordinate graphs with equations in many different relations of priority, depending on the mathematical situation. Our argument here is only that in this situation, and in many school mathematics

situations, equations often take priority whereas they rarely do in work situations like Jake and Evan's.

References

Becker, H. S. (1986). A school is a lousy place to learn anything. In H. Becker (Ed.), *Doing things together: Selected papers,* pp. 121–136. Evanston, IL: Northwestern University Press.

Coulter, J. (1989). *Mind in action.* Atlantic Highlands, NJ: Humanities Press International.

Coulter, J., & Parsons, E. D. (1990). The praxiology of perception, visual orientations, and practical action. *Inquiry, 33* (3), 251–272.

deGroot, A. D. (1965). Thought and choice in chess. In P. C. Wason & P. N. Johnson-Laird (Eds.), *Thinking and reasoning,* pp. 145–151. Middlesex, England: Penguin Books.

diSessa, A. (1991). Epistemological micromodels: The case of coordination and quantities. In J. Montagero & A. Tryphon (Eds.), *Psychologie génétique et sciences cognitives,* pp. 169–194. Geneva: Fondation Archives Jean Piaget.

Garfinkel, H. (1967). *Studies in Ethnomethodology.* Englewood Cliffs, NJ: Prentice-Hall, Inc.

Goodwin, C. (1994). Professional vision. *American Anthropologist, 96* (3), 606–633.

Goodwin, C. (1995). Seeing in depth. *Social Studies of Science, 25* (2), 237–274.

Goodwin, C., & Goodwin, M. (1996). Seeing as situated activity: Formulating planes. In Y. Engestrom & D. Middleton (Eds.), *Cognition and communication at work,* pp. 61–95. New York: Cambridge University Press.

Hall, R. (1995). Realism(s) for learning algebra. In C. B. Campagne, W. Blair, & J. Kaput (Eds.), *The algebra initiative colloquium* (Vol. 1), pp. 33–51. Washington, DC: U.S. Department of Education, Office of Educational Research and Improvement.

Hall, R., & Stevens, R. (1995). Making space: A comparison of mathematical work in school and professional design practices. In S. L. Star (Ed.), *The cultures of computing,* pp. 118–145. Oxford: Blackwell Publishers.

Hall, R., & Stevens, R. (1996). Teaching/learning events in the workplace: A comparative analysis of their organizational and interactional structure. In G. W. Cottrell (Ed.), *Proceedings of the Eighteenth Annual Conference of the Cognitive Science Society,* pp. 160–165. Mahwah, NJ: Erlbaum.

Hanson, F. A. (1993). *Testing testing: Social consequences of the examined life.* Berkeley, CA: University of California Press.

Hanson, N. R. (1969). *Perception and discovery: An introduction to scientific inquiry.* San Francisco, CA: Freeman, Cooper, & Company.

Hutchins, E. (1995). *Cognition in the wild.* Cambridge, MA: MIT Press.

Kuhn, T. S. (1970). *The structure of scientific revolutions.* Chicago: University of Chicago Press.

Latour, B. (1987). *Science in action: How to follow scientists and engineers through society.* Cambridge, MA: Harvard University Press.

Latour, B. (1990). Drawing things together. In M. Lynch & S. Woolgar (Eds.), *Representation in scientific practice,* pp. 19–68. Cambridge, MA: MIT Press.

Latour, B., & Woolgar, S. (1986). *Laboratory life: The construction of scientific facts* (2nd ed.). Princeton, NJ: Princeton University Press.

Lave, J. (1988). *Cognition in practice: Mind, mathematics, and culture in everyday life.* New York: Cambridge University Press.

Lave, J., & Wenger, E. (1991). *Situated learning: Legitimate peripheral participation.* New York: Cambridge University Press.

Lehrer, R., Schauble, L., Carpenter, S., & Penner, D. (1996). *The interrelated development of inscriptions and conceptual understanding.* Unpublished manuscript, University of Wisconsin – Madison.

Lynch, M. (1985). *Art and artifact in laboratory science: A study of shop work and shop talk in a research laboratory.* Boston, MA: Routledge & Kegan Paul.

Lynch, M. (1990). The externalized retina: Selection and mathematization in the visual documentation of objects in the life sciences. In M. Lynch & S. Woolgar (Eds.), *Representation in scientific practice,* pp. 153–186. Cambridge, MA: MIT Press.

McNeill, D. (1992). *Hand and mind: What gestures reveal about thought.* Chicago, IL: University of Chicago Press.

Messer-Davidow, E., Shumway, D. R., & Sylvan, D. J. (1993). Introduction: Disciplinary ways of knowing. In E. Messer-Davidow, D. R. Shumway, & D. J. Sylvan (Eds.), *Knowledges: Historical and critical studies in disciplinarity,* pp. 1–21. Charlottesville, VA: University Press of Virginia.

Michotte, A. (1954/1991). *Michotte's experimental phenomenology of perception.* G. Thines, A. Costall, & G. Butterworth (Eds.). Hillsdale, NJ: Erlbaum.

Newman, S. (in press). *Machinations of the middle: Networked software engineering as lived work and material-semiotic practice.* Unpublished doctoral dissertation, University of California, Berkeley.

Nichols, E. D., Edwards, M. L., Garland, E. H., Hoffman, S. A., Mamary, A., & Palmer, W. F. (1992). *Holt Algebra I.* Austin, TX: Holt, Rinehart and Winston, Inc.

Rogoff, B. (1990). *Apprenticeship in thinking: Cognitive development in social context.* New York: Oxford University Press.

Roth, Wolff-Michael. (1996). Art and artifact of children's designing: A situated cognition perspective. *Journal of the Learning Sciences, 5* (2), 129–166.

Schegloff, E. A. (1984). On some gestures' relation to talk. In J. M. Atkinson & J. Heritage (Eds.), *Structures of social action: Studies in conversation analysis,* pp. 266–296. London: Cambridge University Press.

Schegloff, E. A. (1992). On talk and its institutional occasions. In P. Drew & J. Heritage (Eds.), *Talk at work: Interaction in institutional settings,* pp. 101–134. London: Cambridge University Press.

Schoenfeld, A. H., Smith, J. P., & Arcavi, A. A. (1993). Learning: The microgenetic analysis of one student's understanding of a complex subject matter domain. In R. Glaser (Ed.), *Advances in instructional psychology* (Vol. 4), pp. 55–175. Hillsdale, NJ: Erlbaum.

Schwartzburg, A., & Stevens, R. (1996). *Making space: A collaborative investigation of embodied engineering knowledge.* Berkeley, CA: Hayden Press.

Scribner, S., & Cole, M. (1981). *The psychology of literacy.* Cambridge, MA: Harvard University Press.

Stevens, R. (in press). *Disciplined perception: Comparing the development of embodied mathematical practices at work and at school.* Unpublished doctoral dissertation, University of California, Berkeley.

Strauss, A. (1985). Work and the division of labor. *The Sociological Quarterly, 26* (1), 1–19.

Street, B. (1984). *Literacy in theory and practice.* London: Cambridge University Press.

Suchman, L. (1987). *Plans and situated actions: The problem of human–machine communication.* London: Cambridge University Press.

Traweek, S. (1988). *Beamtimes and lifetimes: The world of high energy physicists.* Cambridge, MA: Harvard University Press.

Wartofsky, M. W. (1979). *Models: Representation and the scientific understanding.* Boston: Reidel.

Wittgenstein, L. (1953). *Philosophical investigations.* New York: Macmillan.

Part II

Teaching Mathematical Talk

Chapters 6, 7, 8, and 9 all study lessons from the same class – the fifth-grade mathematics class from which the example at the beginning of the book is drawn. In chapter 6, Magdalene Lampert provides an overview of the issues involved in studying the teaching of mathematical discourse in school. She describes the investigation of her own practice as a fifth-grade mathematics teacher and the data base which was created around it: records of teaching and learning mathematics for every day across a whole school year. Peggy Rittenhouse, Merrie Blunk, and Peri Weingrad use that data base to analyze the teaching in Lampert's classroom as an intervention in how students talk about mathematics. Finer and finer lenses are applied to tease apart just what the teacher does to model, support, and maintain mathematical discourse.

Rittenhouse reconceives the teacher's role in terms of promoting mathematical "literacy." She calls what the teacher does in that role "stepping in" and "stepping out" to emphasize the dual nature of the teacher's participation in classroom talk. Blunk investigates one element of the teacher's stepping in to classroom discourse. She examines, across the entire school year, what the teacher says to her class about working on mathematics with their peers in small groups. She argues that norms of peer interaction are deliberately established by teacher intervention and, once established, require continuing intervention to be maintained. Weingrad inspects the text of teacher talk through a sociolinguistic microscope, looking for evidence of the respect for and interest in students' ideas that reformers claim is fundamental to learning mathematics. Asking how we would establish that intellectual respect had been taught and learned, she turns to politeness theory and develops analytic tools for coding classroom talk. She focuses particularly on the differentiation between global expressions of respect and interest and those that focus particularly on ideas. Her purpose is not to prove that the teacher has been successful but to investigate a methodology for studying that question that ultimately could support making claims about whether students learn to respect and express interest in other students' ideas.

6 Investigating Teaching Practice

Magdalene Lampert

In this section of the book, three studies of mathematical talk in my fifth-grade mathematics classroom will be presented. The first of these studies, by Peggy Rittenhouse, examines, in her words, "how one teacher helped her students develop ways of talking about mathematics." Using records of what happened during the month of September in my classroom, Rittenhouse examines the teacher's role as participant in and commentator on mathematical discourse, bringing perspectives from current work on literacy learning.

How Does the Teacher "Help" Students Learn to Talk Math?

When Rittenhouse uses the word *helped* to describe what the teacher did, she leaves much to the interpretation of the reader. One ambiguity in the way she formulates her research question lies in the difference between intending to do something and accomplishing it. Does she mean that she is going to describe what the teacher *did* and offer some grounded speculations on the intentions of these actions? Or is she claiming that the teacher was successful in helping her students and then examining evidence for what she accomplished? A second ambiguity revolves around the word *helped* used to describe what the teacher did. As James Gee points out in his afterthoughts to Deborah Hicks's *Discourse, Learning, and Schooling,* our attention to Vygotsky and the importance of the perspective of practice on language learning raises questions about the extent to which the teacher's help should be explicit (1996, p. 269). In order to help, does the teacher explicitly tell students how to talk about mathematics? Or does she engage with them in such talk, scaffolding their efforts and making room for regular practice?

I do not point out these ambiguities here because I think that Rittenhouse should have been more explicit. To the contrary, I find her wording useful both for raising some crucial questions about how we might actually study the relationship between a teacher's intentions and evidence of success, particularly where questions about teaching and learning mathematical talk in school are on the table, and for pointing to some equally fundamental questions about what teaching entails when mathematical discourse is its object.

Rittenhouse clarifies her intention: It is to study what the teacher does and to offer some analysis of the various roles a teacher might play in relation to students' acquisition of mathematical literacy. She borrows from more general scholarship on language learning to posit mathematics as one among many "literacies" students must acquire to function in school. She takes the talk of the first month of a mathematics class as her text and performs an exegesis. She makes sense of the teacher's actions by interpreting them as "stepping in" and "stepping out," participating in and commenting on students' mathematical talk. Her findings are reported as "one teacher's vision of what fostering students' understanding of mathematics might look like."

The question of why we should be interested in this teacher's vision is a complicated one to which I will return. Rittenhouse and (in a later chapter) Merrie Blunk argue that there are few analyses of teaching practice in classrooms where a teacher is explicitly teaching mathematical talk. This may be enough of a reason to pay attention. But as we strive to relate research and reform, we wish for more. We want to be able to say that if a teacher "helps" in a particular way, then students will learn what we want them to learn. Blunk's chapter and Peri Weingrad's are a beginning attempt to examine what it would take to make connections between what a teacher does and what students learn of discourse.

Blunk looks through a wide-angle lens to investigate one element of teacher talk essential to the practice of mathematical talk in school. She investigates, across the entire school year, what the teacher says to her class about working on mathematics with their peers in small groups. She argues that norms of peer interaction are deliberately established by teacher intervention (what Rittenhouse calls "stepping in") and, once established, require continuing intervention to be maintained. The teacher steps in both to execute deliberate plans for how small groups should be structured to support mathematical talk and spontaneously in response to situations that arise every day in the classroom.

Weingrad inspects the text of teacher talk through a sociolinguistic microscope, looking for evidence of the respect for and interest in students' ideas that reformers claim is fundamental to learning mathematics.

Asking how we would establish that intellectual respect had been taught and learned, she turns to politeness theory and develops analytic tools for coding classroom talk. She focuses particularly on the differentiation between global expressions of respect and interest and those that focus particularly on ideas. Her purpose is not to prove that the teacher has been successful, but to investigate a methodology for studying that question that ultimately could support making claims about whether students learn to respect and express interest in other students' ideas. She develops a coding scheme for the teacher's "politeness strategies," claiming that such strategies make it possible for all participants in classroom discourse to cope with the risks entailed in talking about ideas. Making claims about what students learn would additionally entail characterizing the risks and methods of coping that students use and establishing that the use of these strategies increases over time. In order to show such an increase, one would first need to characterize the baseline frequency of related interaction patterns and examine fluctuations against this baseline.

Having conceived of it, we can imagine that such research would be possible although time consuming. That it can even be imagined I attribute at least in part to my attempts, over the past several years, to study my own classroom practice. From the inside, I have been able to appreciate the complexities of doing what Rittenhouse calls "helping my students develop ways of talking about mathematics." I have learned to be wary of simple cause and effect claims about the success of teachers' attempts to "help." This book, and these three chapters in particular, begin to get at the roots of that complexity.

Who Is This Teacher and What Does She Think She Is Doing?

I would like to return to the question of why anyone should be interested in "this teacher's vision of what fostering students' understanding of mathematics might look like." Teaching fifth-grade mathematics for 8 years while doing research on teaching gave me an extended opportunity to articulate and test my vision. Clearly, I am not a typical fifth-grade teacher. The purpose of this research is not to examine what is typical but rather to describe what occurs in the process of attempting to design and implement a particular kind of pedagogical practice.[1] If we are to understand whether and how to implement such practices more broadly, we need to understand them not only conceptually but practically. My classroom was ordinary in the sense that it was located in a midwestern public school, it contained 29 students of varying abilities and varying family backgrounds,[2] and I was expected to abide by the same curriculum as everyone else in the district. In an odd sense, I was more "ordinary" than many of the teachers who are written about in this book in that my practice was not guided by a particular model of curriculum or instruction derived from an identifiable theory or research finding. Like many teachers, I read a lot of theory and research findings, and my practice is informed by what I read, but I do not seek to *implement* or *apply* theory and research in the sense those terms are often used.[3]

I have been curious for a long time about what it might mean to do and know mathematics. I began teaching mathematics in 1969 as a newly minted Bachelor of Science. I had majored in mathematics as an undergraduate and took no teacher preparation courses. But I did study "mathematical talk" in the course of doing a thesis in the philosophy of mathematics, and I "talked" mathematics in most of my courses. In 1982, I began teaching fifth-grade mathematics concurrently with conducting research on my own teaching. Until 1989, I worked independently on designing and analyzing my practice. At that point, I started to build a research team who could collect more extensive records in my classroom than I could by myself and analyze those records from various theoretical perspectives.

When I began to teach fifth grade, I self-consciously set out to design a pedagogy that would engage students in mathematical work of the sort I had learned about in college. I began to experiment with various elements of my practice, using ideas from mathematical philosophy and social science to design promising approaches to teaching and learning authentic mathematics. For example, drawing on the work of Georg Pólya (1954) and Imre Lakatos (1976) on mathematical discourse, I began to organize whole class discussions around students' different answers to a problem, and to call their answers "conjectures" until the class as a whole discussed the legitimacy of various solution strategies. In these discussions, the mathematical importance of conditions, assumptions, and interpretations was made more explicit, and students were pushed toward revising definitions of their own terms of reference. Because this activity was conducted in a classroom, it required additional work from teacher and students to deflect the agenda from the simple production and assessment of "the answer" so that students could learn the mathematical processes of refining language and verifying assertions. Instruction had to be slowed down, and the discourse opened up to enable students to participate in evaluating their own thinking.

Another example of the sort of design experiment I have undertaken, drawing on work in both cognitive science and mathematical theory, was the use of multiple representations for place value in decimal numbers drawn from students' own attempts to express their thinking in graphic forms. The emphasis in class discussions during these lessons was on students' explaining their conclusions about how to order lists of numbers using more than one representation. This approach involved the teacher and the students in developing new ways of communicating about ideas and required negotiations between school conventions and mathematically appropriate forms of expression. Trying these approaches out in my own daily lessons across several school years, documenting the teaching and learning practices that were needed to support them in that setting, and beginning to develop ways to write about teaching and learning from the perspective of practice have become the components of a new kind of educational research.

I conjecture that much can be learned from examining one attempt to construct practice in light of contemporary scholarship in mathematics, learning, and teaching, for a study of practice will expose issues and reveal problems that could not be known from simply synthesizing theories and research findings from relevant disciplines. This conjecture is based on analogies with work in other fields in which there is an interaction between the use of theory and the design of practice. In order to learn about the practice of architectural engineering, for example, it is considered useful to begin with the study of an existing structure (i.e., one that "works"), take it apart analytically into its elements, model each of them in turn in a less complex framework, and finally examine how all of the elements go together to make the building stand up.[4] At the level of theoretical modeling, the real structure can be understood in relation to the physics of forces and materials. But at the practical level, it can only be understood as the result of the constructed practices that make use of those theories and bring them to bear on an actual building project which is always messier than could be predicted by the theory. Designing a practice extends the process of understanding by allowing one to inquire about the ways in which actions are related to understanding in real time and to critically examine the applicability of ideas. The study of practice, as I have portrayed it here, has both scholarly and practical purposes; it is intended both to construct a theory of practice and to inform practitioners as they do their work.

Research Methods and Procedures

From 1984 to 1989, I collected various kinds of records of my own teaching practice and the learning practices of my students. During

1989–90, I further collected and electronically catalogued a large multimedia corpus of records of what was happening in my classroom.[5] These records include videotapes of large and small group work for nearly every day of the year, audio transcripts of many lessons, observers' notes on all lessons, all of the students' daily written work for the year, my own journal of plans for and reflections on all lessons, and interviews with teacher and students before, during, and after the school year. The records of teaching and learning in my classroom in 1989–90 were collected according to guidelines drawn from ethnographic research methodology. In order to collect video and audio that adequately represented classroom events, cameras and microphones were placed in different locations and the tape ran constantly from before the lesson began until a few minutes into the next lesson or event.[6] Daily written documentation of classroom events during the year of videotaping was produced by trained observers using a format developed by Deborah Ball and myself. Records of lessons were prepared by four different observers each week, helping to make individual biases and preconceptions in what the observers chose to record more explicit and providing a range of interpretations of events. Having such an extensive collection of information about what occurred in the classroom across the year to share with a team of researchers and consultants has made it possible for the research team to look at what occurred from many different points of view.[7]

Since the availability of such records and annotations has been uncommon, appropriate research procedures for using the classroom records have had to be created and adapted from various more conventional approaches to studying teaching and learning. How do records of practice become data for research? What research procedures are appropriate to identify the elements of this practice, to explicate their contents, and to relate them? How can the records of a whole year of teaching and learning be analyzed in a way that respects the interconnectedness of aspects of classroom life, while at the same time enabling the study of distinct parts of it? How can we communicate the idea that the whole year is a "case" of mathematics teaching and learning, while focusing on single interactions, lessons, or groups of lessons? How can we check on the inferences we are making about how the data represents aspects of mathematics or understanding or teaching? And finally, how can reports on this work be produced that represent the particular dynamics of the situation under study, while also having something to say to researchers and practitioners about a theory of mathematical pedagogy? The three studies we present here are a partial effort to work on these methodological puzzles.

Tools and Strategies for Analysis

The records of teaching and learning on which this study is based have been catalogued in an electronic data base. Access to these data sources and their subsequent analysis by a research team has been enabled through the development of software that makes it possible to browse, mark, link, and annotate the materials.[8] The production of electronic tools for searching, identifying, and linking information has enabled a research team to describe and analyze what the pedagogical activities of establishing a culture of inquiry, identifying and using new modes of communication, and restructuring the curriculum around authentic mathematical activities are like in an actual classroom setting and what sort of practices of teaching and learning they entail. These tools also facilitated the creation of new categories for analysis as the study progressed.

In order to examine the practical dynamic elements of establishing and maintaining a culture, developing and using tools for mathematical communication, and creating a curriculum in the context of work on problems, we have searched the records for data that would enable us to study four kinds of interactions: between teacher and student(s), between student(s) and student(s), between teacher and subject matter, and between student(s) and subject matter. While it is understood that these interactions do not occur in isolation from one another, they can be analyzed separately to get at the contents of elements of mathematics teaching and learning, and they then can be integrated into a picture that represents the classroom culture as a whole in terms of connections among teacher, students, and mathematics. Particular interactions can initially be studied in one kind of data or another, but they will also be manifest in other kinds of documentation, so evidence drawn from one domain can be compared with interpretations of what is occurring in other domains. For example, students' interactions with a particular mathematical idea might most directly be ascertained from an analysis of what they write in their notebooks, but evidence about this interaction can also be gained from the ways in which they participate in whole-class teacher-led discussions and small group discussions with their peers in which that same mathematical idea is prominent. These multiple sources of data are also a protection against biases in interpretation, as their use enables triangulation among different kinds of information about the same phenomenon (Webb, Campbell, Schwartz, & Sechrest, 1966).

To examine social interaction and the interaction between persons and ideas, the research team has used methods and categories developed by sociolinguists and interaction analysts to study social competence. These methods support the study of key elements of communication and the role of the social setting in shaping individual activity.[9] Because the kind of social competence we wanted to study is mathematical, we have created analytic categories with themes that emerge from a consideration of disciplinary processes and content. Relevant sites for analysis have included the physical and linguistic determination of different kinds of contexts and tasks, teacher and student roles and how they are constructed in different kinds of participation structures, and the mechanisms that determine transitions from one kind of activity structure to another (see Florio-Ruane, 1987; Gumperz, 1976). Students' and teachers' attempts to communicate with one another using spoken language, writing, and graphic representations have been examined, both for their content and for what they represent about the kind of discourse that is occurring.

The research studies included here by Rittenhouse, Weingrad, and Blunk are a subset of the reports that have been produced by this project. The method of analysis we have employed in the following three studies is something like "exegesis," where video and audiotapes, the teachers' and students' notebooks, and the observers' notes are taken to be a text with multiple levels of meaning – a "tangled web," as Clifford Geertz (1973) calls it, that needs to be both unraveled and respected for the complexity it represents. We assume that like all human action, teaching does not have simple explanations, that there are many ways to interpret any action, and that the levels of meaning that can be found are confounded and sometimes in conflict. In order to communicate about teaching in ways that retain the dynamic quality of the practice, such a methodology, though messy, seems appropriate. The purpose of such interpretive research is not to determine whether general propositions about learning and teaching are true or false, but to further our understanding of these particular kinds of human activity in the contexts where they occur.

Notes

1. Other such efforts include Schools for Tomorrow and QUASAR. See Brown (1992), Silver, Smith, and Nelson (1995), and Silver and Stein (1996).
2. Although the school where I taught was near a university, the students in the school were not children of university faculty.
3. See Lampert (in press).
4. See Salvadori and Tempel (1983). Joseph Schwab (1978) describes a similar process in his analysis of theories of practice in education.
5. This project (Mathematics and Teaching through Hypermedia or M.A.T.H.), funded by the National Science Foundation, collected information about teaching and learning

in two classrooms where new approaches to mathematics education were being used to give prospective and practicing teachers and teacher educators images of what a different kind of practice could look like. The M.A.T.H. Project was codirected by Deborah Ball and myself. See Lampert and Ball (in press) for an extended description of this project.

6. See Erickson and Wilson (1982) for explicit guidelines for videotaping based on principles of ethnographic research.
7. The use of these records to conduct research on teaching, including the studies reported in chapters 7, 8, and 9 of this book, was sponsored by the Spencer Foundation.
8. Software to carry out these actions was developed by Mark Rosenberg and Jim Merz.
9. In the examination of the data and the creation of analytic frameworks, we follow procedures similar to those outlined by Erickson and Shultz (1982) for determining the constituent structure of social occasions.

References

Brown, A. (1992). Design experiments: Theoretical and methodological challenges in creating complex interventions in classroom settings. *Journal of the Learning Sciences, 2* (2), 141–178.

Erickson, F., & Shultz, J. (1982). When is a context? Some issues and methods in the analysis of social competence. *Quarterly Newsletter of the Laboratory for Comparative Human Development, 1* (1), 5–10.

Erickson, F., & Wilson, J. (1982). *Sights and sounds of life in schools: A resource guide to film and videotape for research and education* (Research Series No. 125). East Lansing, MI: Institute for Research on Teaching.

Florio-Ruane, S. (1987). Sociolinguistics for educational researchers. *American Educational Research Journal, 24,* 185–197.

Gee, J. (1996). Vygotsky and current debates in education. Some dilemmas as afterthoughts to *Discourse, language, and schooling.* In D. Hicks (Ed.), *Discourse, language, and schooling,* pp. 269–282. New York: Cambridge University Press.

Geertz, C. (1973). *Interpretation of cultures.* New York: Basic Books.

Gumperz, J. J. (1976). Language, communication, and public negotiation. In P. Sanday (Ed.), *Anthropology and the public interest: Field work and theory,* pp. 273–292. New York: Academic Press.

Lakatos, I. (1976). *Proofs and refutations.* New York: Cambridge University Press.

Lampert, M. (in press). Teaching as a thinking practice. In J. Greeno & S. Goldberg (Eds.), *Thinking practices.* Hillsdale, NJ: Erlbaum.

Lampert, M., & Ball, D. L. (in press). *Teaching, multimedia and mathematics: Investigations of real practice.* New York: Teachers College Press.

Pólya, G. (1954). *Mathematics and plausible reasoning* (Vol. 1 & 2). Princeton, NJ: Princeton University Press.

Salvadori, M., & Tempel, M. (1983). *Architecture and engineering: An illustrated teacher's manual on why buildings stand up.* New York: New York Academy of Sciences.

Schwab, J. (1978). The practical: Arts of eclectic. In I. Westbury & N. Wilkof (Eds.), *Science, curriculum, and liberal education: Selected essays,* pp. 322–366. Chicago: University of Chicago Press.

Silver, E. A., Smith, M. S., & Nelson, B. S. (1995). The QUASAR project: Equity concerns meet mathematics education reform in the middle school. In W. G.

Secada, E. Fennema, & L. B. Adajian (Eds.), *New directions in equity and mathematics education,* pp. 9–56. New York: Cambridge University Press.

Silver, E. A., & Stein, M. K. (1996). The QUASAR project: The "revolution of the possible" in mathematics instructional reform in urban middle schools. *Urban Education, 30* (4), 476–521.

Webb, E. J., Campbell, D. T., Schwartz, R. D., & Sechrest, L. (1966). *Unobtrusive measures: Nonreactive research in the social sciences.* Chicago: Rand McNally.

7 The Teacher's Role in Mathematical Conversation: Stepping In and Stepping Out

Peggy S. Rittenhouse

This chapter is about the role of the teacher in helping students learn mathematics. While that statement may sound rather straightforward, it isn't. That's because what I mean by mathematics may not be the same as what others mean when they think "mathematics." So, before I discuss the teacher's role in mathematics learning, I'd like to discuss what I mean by mathematics. In order to do that, I'm going to back up even further and discuss a term that may at first seem unrelated to mathematics: literacy. As I hope will soon be clear, understanding literacy is fundamental to understanding mathematics, as well as the role of the teacher in mathematics learning.

Understanding Literacy

When many people think about literacy, they think "reading and writing." This seems logical; after all, variations of this definition have been with us since the Middle Ages (Venezky, 1991). But defining literacy as "the ability to read and write" (*American Heritage Dictionary,* 1983) becomes problematic when literacy situations are examined. Does literacy include oral language, as when a group of people discuss a written text, or is literacy limited to an individual's silent interaction with written language? Does literacy deal only with issues related to those disciplines associated with reading and writing, or does literacy encompass other disciplines like science and mathematics? After all, people *do* read and write about issues like chaos theory, fractals, and the 25%-off sale at the local mall. In short, "the ability to read and write" provides us with, at best, a fuzzy picture of literacy. In recognition of this, those interested in literacy have tried to contextualize literacy. The result has been a veritable stew of literacy terms: "inert" (Cremin, cited in Venezky, 1991, p. 49), indicating a marginal level of literacy; "functional" (Stedman & Kaestle, 1991) or "liberating" (Cremin, cited in Venezky, 1991,

p. 49), indicative of the level of literacy needed by an individual to do what needs to be done to function in everyday life; and "critical" (McLeod, 1986), which acknowledges a person's ability to use literacy to act on the world. All these terms hint strongly of the social nature of literacy and help us understand that literacy is not (only) a quality of the individual; it is also embedded in social contexts, making literacy a dynamic concept. As Scribner (1984) points out, individuals do not

> extract the meaning of written symbols through personal interaction with the physical objects that embody them. Literacy abilities are acquired by individuals only in the course of participation in socially organized activities with written language. . . . It follows that individual literacy is relative to social literacy. Since social literacy practices vary in time (Resnick [1983] contains historical studies) and space (anthropological studies are in Goody [1968]), what qualifies as individual literacy varies with them. . . . Literacy has neither a static nor a universal essence. (pp. 7, 8)

This means literacy, and being literate, is different in different contexts. Gee (1991) acknowledges this in his discussion of discourse. According to Gee, "discourse" describes the particular ways in which language, thoughts, and actions are used by members of particular groups. Thus, if you are a member of a "teenage group," you will not only dress and act, but also talk, in ways that are different from other groups, like "teacher." While people can (and do) belong to more than one discourse group, they must remember to dress, act, and communicate in ways that are appropriate for the group currently being joined. To do otherwise would mark a person as "not a member" of that particular group.

For Gee, literacy is the ability to control the language used in discourses other than the primary discourse within which a person is raised. According to Gee, total mastery of these "secondary discourses" (1991, p. 8) requires opportunities to both *acquire* that discourse through immersion in the social aspects of the discourse (e.g., ways of acting and believing associated with the discourse), as well as by *learning* the structural aspects of the discourse (e.g., the grammar, mechanics, and style of the discourse). Thus, a discourse is an "identity kit" (1991, p. 1) consisting of not only the language of the discourse but also its ways of acting and thinking about the world.

The view of literacy on which the analysis in this chapter rests draws heavily on Gee's ideas, but it also includes notions of emergent and critical literacy as well. In short, I define literacy as

> the continual development of the ability to engage in a variety of discourses, combined with an ongoing reflection about the (re)positioning of those discourses relative to one

another, in order to actively view, question, and use the strengths and weaknesses of different discourses as a means of enhancing one's fluent control of those discourses.

In other words, literacy is a lifelong endeavor in which individuals are differently literate depending on the context. Thus, a person enrolled in a teacher education program at a university might be highly literate as a "college student" (able to speak, act, dress, and think in ways that clearly mark one as a member of the group "college students") and less literate as a "teacher" (because that individual is just beginning to acquire and learn the discourse of "teacher"). So, what does all of this have to do with mathematics?

Understanding Mathematical Literacy

Mathematics, like other areas of human endeavor, has its own vocabulary and ways of thinking about the world that mark it as "mathematics." People who engage in mathematics thus participate in mathematics discourse communities. Even within the field of mathematics, highly specialized discourse communities exist, each using language and ways of thinking that differ from other mathematics discourse communities (Euclidean geometry is different from non-Euclidean geometry, for example). The mathematics most schoolchildren in the United States encounter is an example of yet another mathematics discourse community. Traditionally, this discourse community has been marked by its heavy emphasis on algorithms, push to get "the right answer," and a lack of talk about mathematics among student members. This focus on the nuts and bolts of mathematics is similar to a focus on grammar and mechanics in writing. In order to communicate their ideas effectively, people do need to use those conventions the intended audience expects. However, emphasizing grammar and mechanics over *what* is said and *how* it is said falls short of the goal of effective communication. In the field of mathematics, this means understanding that while mathematical terms and arithmetic operations, like grammar and writing mechanics, are important, they are only a small part of the whole. They are not the whole of mathematics. Thus, while members of a mathematics discourse community do need to have at their disposal basic tools of the discourse in order to communicate mathematical ideas, they also need to understand how to talk, think, and act mathematically in different contexts. Figure 7.1 shows the similarities and differences in discourse tools among various mathematics discourse communities.

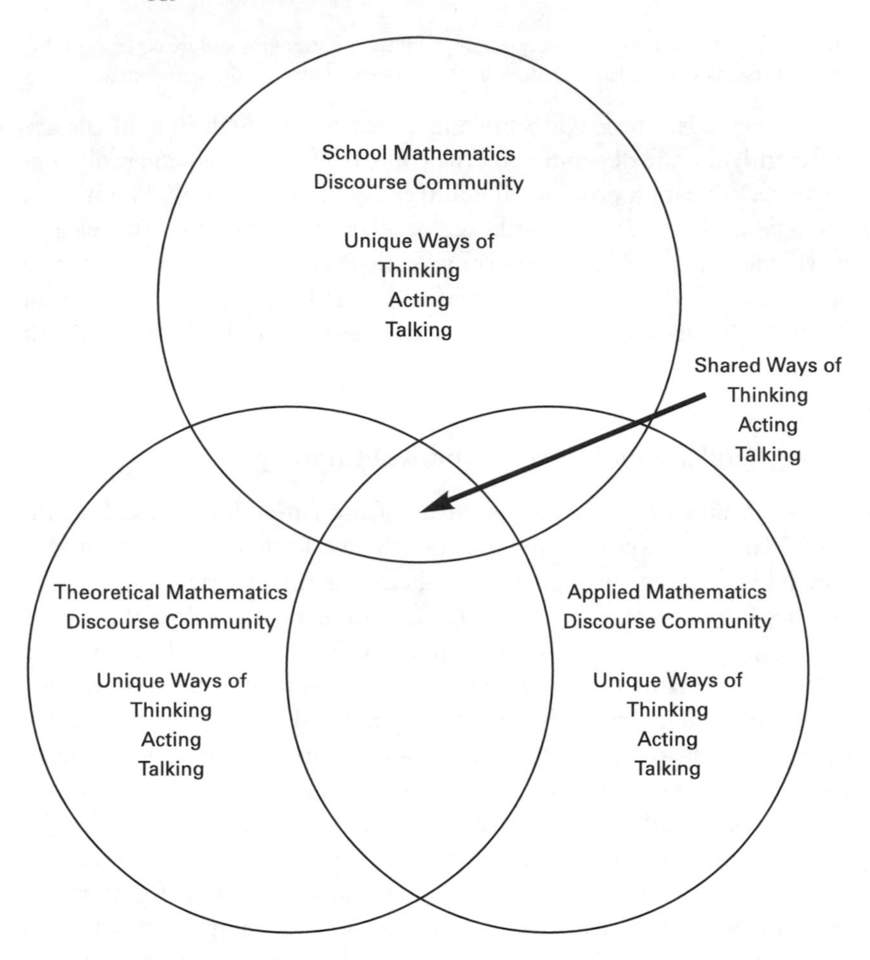

Figure 7.1. Similarities and differences among various mathematics discourse communities

While the discourse communities of theoretical or applied mathematics are also interested in "the right answer," they are much more than this. They also involve ways of speaking and thinking and norms for argument and mathematical proof. If children are to develop a more complete mathematical literacy, they need opportunities to acquire and learn aspects of mathematics discourse communities different from the school discourse. This is not to say fifth graders have to study discrete mathematics; obviously they don't. But children do need opportunities to move in and out of different mathematics discourses. Doing so provides

Figure 7.2. Movement of school mathematics discourse into and out of other mathematical discourses

students standpoints above all the discourses from which to view the strengths and weaknesses of each. Understanding the various strengths and weaknesses of these different mathematics discourses can help students develop their fluent control over them (as Figure 7.2 shows) and, ultimately, enhance their mathematical literacy.

Talk As a Way to Learn Mathematics Content

Talk is the primary vehicle students use to move between different mathematics discourses. It is what allows students to view the strengths and weaknesses of their school mathematics discourse and other mathematics discourses; ultimately, talk is the vehicle that helps students enhance their mathematical literacy or what the National Council of Teachers of Mathematics (NCTM) calls their *mathematical power*. NCTM defines this as "an individual's abilities to explore, conjecture, and reason logically, as well as the ability to use a variety of mathematical methods effectively to solve nonroutine problems" (1989, p. 5). Central to this notion of power is the idea that students will have a more complex and connected understanding of mathematics, that they will come to view mathematics as more than the sum of its conceptual and algorithmic parts. Students who develop this kind of power will have at their command ways of using mathematical language, ways of thinking about mathematics, and ways of acting mathematically that will enable them to engage in a wide range of mathematical problem solving (Gee, 1991). Developing this kind of mathematical power requires teachers and students to engage in talk with one another about mathematical problems and problem solving (NCTM, 1991). Yet instruction centered around talk about mathematics is not often found in American classrooms.[1]

Traditionally, teachers have been viewed as the sole classroom authorities about mathematics. They decided which math content was to be learned, they demonstrated how to solve problems, and they evaluated students' responses to problems. Students, in contrast, listened to their teachers explain how to do problems, and then they worked individually (and quietly) to solve various problem sets. As a result of this kind of instruction, teachers were generally the only persons in the room who actually talked about mathematics. But there is a difficulty with this view of instruction wherein teachers tell students what problems to work on and how to do them. When teachers transmit, and monitor for the correct reception of, a body of knowledge, it can appear that mathematics is a field made up of stable and well-defined knowledge (Brown, Cooney, & Jones, 1990), and this can lead to a belief that what is known about mathematics was established long ago. One consequence of this view of mathematics is that both teachers and students are likely to see themselves as passive receivers of others' knowledge.

In contrast to what many nonmathematicians believe about mathematical knowledge, those in the field tell us mathematical knowledge is not fixed; rather, like other knowledge, mathematical knowledge is socially constructed and can (and does) change over time (Kitcher, 1986). From this perspective, mathematics is much more than a series of algorithms to be mastered or facts to be passed from generation to generation. It is also a way of talking about mathematics in order to expand what we know, a way of using specialized procedures for proving and disproving ideas in order to build consensus in the field (Lakatos, 1976).

How does all of this connect to "literacy" and "discourse"? The reforms outlined above suggest that students need more than the traditional kinds of mathematics instruction found in school. Further, the reforms suggest talk as a vehicle for extending students' learning about mathematics. Like those who talk about discourses and discourse communities, those interested in mathematics education reform are also calling for students to extend their understanding of mathematics as more than algorithms. Whether the result is called mathematical literacy or mathematical power, the point is that talk plays a central role in helping students better understand mathematics. But how can we help students talk about mathematics when they likely have had few opportunities to do so? One way is by looking at what teachers do to help their students learn to talk about mathematics. In this chapter, I examine how one teacher helped her students develop ways of talking about mathematics. Understanding how she did this and the role her talk played in helping those students learn how to talk mathematically may be helpful for others interested in fostering students' development of mathematical power. In my discussion, I draw on transcripts of classroom conversations taken from the first month of school during the 1989–90 school year. I have chosen to look at this particular time because it was then that Lampert and her students first began developing the kind of conversational norms and routines they would use throughout the school year.

Why Look at Teacher Talk?

Lampert uses a variety of strategies to support her students' growth as mathematical knowers, including tasks, routines, and providing opportunities for talk.[2] While selecting appropriate tasks, establishing supportive routines, and creating opportunities for talk are important, they are not enough. Students do not automatically begin talking about mathematics in meaningful ways simply because they are presented with appropriate tasks or are placed together in groups and told, "Talk to each other." They must learn how to talk with one another about mathematics; they need to learn a mathematical discourse.

Back to Gee: Or What Is a Discourse and Where Do I Get One?

According to Gee, a discourse can be thought of as an "identity kit" (1991, p. 3) made up of ways of thinking, acting, and speaking. The kind of talk used within a particular discourse, for example, allows one to be recognized by other members of that discourse group as a member. Thus, discourse is not only a broad, general term, but one that is also quite specific, depending on the situation. The discourse of mathematicians, for example, is different from the discourse of attorneys in part because the kinds of talk and ways of thinking about the world associated with each group are different. For Lampert's students, talking about mathematics means using a discourse similar to that of mathematicians. How does one develop the ability to use a particular discourse?

According to Gee (1991), one way is to *acquire* it informally through authentic, ongoing exposure to talk and ways of being. Most infants and young children learn their native language in this way as they hear and are engaged in conversations with those around them. A second way (Gee, 1991) to develop the ability to use a particular discourse is to *learn* it in more formal ways, as is done in school. Systematically studying a particular discourse and identifying its norms, routines, and underlying rules allows one to gain a deeper understanding of the ways in which a particular discourse functions, providing an individual the tools to use that discourse effectively. So, for Lampert's fifth graders, talking about mathematics provides them opportunities to both acquire and learn a particular mathematical discourse and enhances their development of "mathematical power" (NCTM, 1989).

So, it would seem that talking about mathematics is "a good thing." Exposure to mathematical talk, however, does not guarantee that students will automatically understand and use this kind of talk. Nor do opportunities for authentic talk about mathematics ensure that students will come to understand the norms and rules of mathematical discourse. In order to learn the structure of mathematical discourse, students need opportunities to learn how the discourse they are using works; they need to "talk about the talk." Only then can they develop metalinguistic power over the discourse. The kind of talk teachers use with their students can help students both acquire and learn a new discourse.

What's a Teacher to Do?: Comprehensible Input and Mathematical Discourse

In order for students to learn about the structure of the discourse of mathematics – its rules for posing and providing evidence for conjectures, ways of arguing about ideas, and so forth – they need opportunities to stop and think about that structure. Teachers can help students do this using talk designed to point out features of classroom conversations that are representative of mathematical discourse and by providing students information about mathematical vocabulary or ways of presenting ideas that may be new to them. Pointing out features of classroom conversations or providing students new information that can help them make sense of the mathematical talk in classrooms is important for a couple of reasons. First, students who have had few opportunities to engage in talk about mathematics might be overwhelmed by conversations different from those to which they are accustomed. Second, even though some students might wish to engage in mathematical conversations, they may not have the tools necessary to do so. In other words, even when immersed in talk about mathematics, students may not know how to jump into this kind of conversation or how to say what they are thinking. Why so?

Think of mathematical discourse as a "new language" to be learned. For novice users of this new language, it is often very difficult to make sense of what is being said if the conversation "flies by" at normal conversational rates of speed.[3] However, if those who already speak the new language slow down, repeat, or rephrase what they have said, or try to provide novices with other forms of sense-making assistance, the likelihood that novices can understand at least some of the conversation increases. That part of the language novices can pick up on and make sense of within this swirling whole is what Krashen (1977, 1982, 1985) calls "comprehensible input."

The comparison of mathematical discourse to a new language is not a perfect one. After all, many students in U.S. classrooms are speakers of English. However, the way language is used, the specialized terminology, and the style of argument used in reasoning mathematically may be very different from the kinds of talk students expect to engage in at school. So, in classrooms where teachers are trying to help students gain control over a new mathematical discourse, part of the teachers' role is to help students see and understand those aspects of mathematical talk that are "new." One way teachers can try to increase the amount of comprehensible input available to students is by stopping the conversation to point out instances where students are using (or closely approximating) the kinds of talk associated with this discourse. When teachers do this, they "slow down the action" so novice members of the mathematics discourse community have a better chance of understanding what is happening. Teachers can also increase the likelihood their students understand the discourse by explicitly teaching the vocabulary, rules, and conversational norms associated with the new discourse.

As students understand more and more of what is being said, they can begin to acquire and learn about the new discourse and can exercise more control over it in their own mathematical talk. The teacher's role in fostering mathematical discourse among students is one of helping them comprehend and use the discourse to deepen their understanding of mathematics. In other words, the role of teacher talk in developing students' mathematical power is one of supporting students as they move from peripheral participation within the classroom's mathematical discourse community (e.g., observing others engage in the range of talk and action associated with the discourse community) to a more full participation as they engage in a variety of mathematical actions and talk within the classroom discourse community (Lave & Wenger, 1991).

Apprenticeships and Participation

Traditional assumptions about the nature of cognitive development are being reexamined as more is learned about the social nature of learning and the ways in which actions and interactions with one's peers can influence an individual's learning.[4] A useful way to think about the interconnections between learners, actions, and the situated nature of what they do is to view this social milieu as an apprenticeship. Rogoff (1990) describes this apprenticeship model as one that

> has the value of including more people than a single expert and a single novice; the apprenticeship system often involves a group of novices (peers) who serve as resources for one another in exploring the new domain and aiding and challenging one another. Among themselves, the novices are likely to differ usefully in expertise as well. The "master," or expert, is relatively more skilled than the novices, with a broader vision of the important features of the culturally valued activity. However, the expert too is still developing breadth and depth of skill and understanding in the process of carrying out the activity and guiding others in it. Hence the model provided by apprenticeship is one of active learners in a community of people who support, challenge, and guide novices as they increasingly participate in skilled, valued sociocultural activity. (p. 39)

Thus, apprenticeships can provide all members of a community with opportunities to learn and develop; however, as Rogoff points out, not all members will have the same degree of expertise. The degree to which any one individual participates within the community will likely vary, but this variability need not hinder learning. As Lave (1988, p. 2) suggests, "Apprentices learn to think, argue, act, and interact in increasingly knowledgeable ways with people who do something well, by doing it with them as legitimate, peripheral participants."[5] Thus, novices with differing levels of expertise can still participate in and learn from community problem-solving encounters. In fact, it is through their increasing involvement with more knowledgeable peers and teachers that novices move toward full participation as members of the community. In the remainder of this chapter, I examine the ways in which Lampert helped support her students' efforts to participate more fully in the problem solving and talk about mathematics. I do so by examining ways in which Lampert's talk helped students both acquire and learn a mathematical discourse designed to foster their understanding of mathematics, increase their mathematical power, and move toward full participation within this particular mathematical community.

Stepping In and Stepping Out: Lampert's Multiple Roles

As she engaged in mathematical discussions with her students, Lampert shifted back and forth between two roles: that of a participant in the discussion and that of a commentator about the discussion. Each of these roles played a part in helping Lampert's students gain control over the mathematics discourse. As a participant, Lampert "stepped into" authentic mathematical discussions with her students in order to help them acquire competence in the discourse. She listened to students' ideas and asked questions about those ideas when she did not fully understand what a student was trying to say. She contributed to discussions and provided insights based on her own mathematical knowledge as well. Her questions and comments varied in kind and purpose and extended beyond the usual evaluative nature of teacher talk (Mehan, 1979) found in many classrooms. In short, as a participant, Lampert's contributions to class discussions were more conversational than didactic in nature. As such, the ways in which Lampert interacted as a conversational participant with her students provided these novice members legitimate opportunities to interact with one another around mathematics in real ways. That is, despite their limited expertise, these novices were able to grapple with real mathematics issues and participate as real members of the mathematics discourse community. As a result, the kinds of conversation that arose provided all students an opportunity to better "see" mathematical discourse in action.

As a commentator, Lampert "stepped out of" classroom discussions to comment on and more formally teach the rules and norms of the discourse she wished her students to use as they talked about mathematics. She recognized that she was the most knowledgeable person in the room when it came to talking about mathematical concepts and was able to anticipate points in the conversation that might need to be slowed down or "rewound and replayed" so that the less expert members of the discourse community could comprehend what was going on. When students made comments that represented the kind of talk she wished the class as a whole to use, Lampert would point out those instances and tell why they were examples of the discourse. When disagreements broke out among students, she paused to explain the rules of polite argument for mathematical discourse. At times, she "named" mathematical concepts students were describing, thus giving the class the "tools of the trade" they needed to talk about mathematics in ways that more closely resembled the discourse found within the discipline at large. As a commentator, Lampert's contributions were more didactic than conversational in nature, designed to help her students learn about mathematical discourse. More importantly, as a commentator, Lampert's role was to anticipate where the bumps in the conversational road might be and to slow it down so the conversation could be comprehensible for her students. Further, as a commentator, Lampert's role included keeping a close watch on where a student might be headed with an idea, anticipate when the student might not have the expertise to talk about that idea, and provide the kind of support that would allow the student to successfully talk about the idea with the class.

Taken together, Lampert's two roles helped transform the swirling whole of the classroom talk about mathematics in ways that made it more comprehensible for students. Although I will describe Lampert's actions in terms of "stepping in" and "stepping out," the two roles were not clearly separable, nor easily defined, within the context of a particular classroom discussion. Multiple purposes were often present as Lampert participated in and listened to the fifth graders' mathematical conversations, but in order to better understand what her two roles "looked like" in those conversations, I have chosen to discuss them separately. As I discuss each of Lampert's roles, I will provide examples of classroom transcripts to illustrate how she "stepped in" and "stepped out." Keep in mind, however, that these examples do simply illustrate points I am making. They are taken out of context and, as in most teaching situations, more is going on in each segment than I will talk about. As she talked with students, Lampert had to manage multiple, and sometimes conflicting, goals.[6] My purpose in sharing transcript segments is to focus on the interaction of Lampert's conversations with students about mathematics, of learning to say clearly what you mean when you talk about mathematics. Remember, too, that these transcripts are of novices learning a new discourse. The kinds of conversations Lampert and her students engaged in during the first month of school represent "learning how" to talk about mathematics; they are not wholly about doing mathematics.

In short, these early conversations are preliminary ones; they are about learning to say clearly what you mean.

Participating in Mathematical Conversations: Modeling and Scaffolding

As she participated in conversations with her students, Lampert worked to make the conversation more comprehensible for them. She did so by modeling the kind of talk she wished her students to engage in and by scaffolding their efforts to engage in talk that more closely resembled the discourse of mathematicians (Cazden, 1983). In short, as a participant, Lampert used, and helped her students use, the kind of talk needed to discuss their mathematical ideas. In the section below, I analyze a segment of conversation to show how Lampert modeled norms for interaction appropriate to a mathematics classroom as she sought clarification of one student's ideas. Following that, I analyze Lampert's interaction with two students to show how she scaffolded their talk so that the students could work together, helping each other clarify a mathematical idea.

Modeling Requests for Information

In the following segment, which occurred during the first week of instruction, Lampert and students were discussing materials they would use during math class. Each student had been provided a notebook filled with pages of graph paper, and Lampert had introduced the terms *horizontal* and *vertical* in reference to the graph paper lines. She then asked students to share their strategies for remembering to which lines the terms referred, and Shahroukh offered his.

Shahroukh: Usually in different kinds of math, first you use subtraction and then you use addition, so you can think of is first to make a subtraction sign and then put a line, that would be horizontal and see then you go from the alphabet the first word with the first letter in the alphabet, you know "h" is the first one so you know that's a minus and then vertical is the line going down you know because it's "v" because the first letter goes across and then the . . .

Lampert: Okay, so "h" comes before "v" in the alphabet. I'm not sure I know what you mean when you say subtraction comes before addition in math.

(Transcript, 9/11/89)

After Shahroukh offered his explanation, Lampert asked him to clarify what he meant. In so doing, she modeled the conversational norm of politeness. By politeness, I mean using talk in ways that are not deliberately hurtful and that do not work to shut down conversation. If Lampert's students were to develop their mathematical power, they needed opportunities to challenge and clarify one another's ideas, yet doing so could make students feel their persons, rather than their ideas, were being challenged. Lampert's comments modeled the norm of polite interaction, showing students how to talk with one another in ways that reduced the threat to each other's self-esteem.

Shahroukh's comments might at first have seemed unrelated to the discussion at hand. The class was, after all, talking about graph lines, not mathematical operations. Lampert's initial response, "Okay, so 'h' comes before 'v' in the alphabet," acknowledged that she was listening, but her next statement, "I'm not sure I know what you mean when you say subtraction comes before addition in math," indicated she wasn't sure what to make of Shahroukh's ideas. Rather than dismiss what he said and move on to someone else, though, Lampert let Shahroukh know that she was interested in his idea and wanted to know more. How did she do this? By beginning her comment with "I'm not sure I know what you mean," Lampert told Shahroukh, and the rest of the class, that she assumed *he* understood what he was talking about ("I'm not sure I know *what you mean,*" emphasis added), but that *she* needed more information ("*I'm not sure I know* what you mean," emphasis added) in order to share that understanding with him. Her word choice reduced the threat to Shahroukh personally because she depicted herself as the one who did not know. Phrasing her comments in this way kept the conversation open, allowing Shahroukh to clarify, rather than justify, his idea.

Shahroukh: In some sort of math, you know, like I don't know what you call it, when you use brackets and have to use signs, division, those things? They have a way that they say "always try to use subtraction first."

Lampert: Okay, so one term that people use for that math is the order of operations. Subtraction, addition, division, and multiplication. So if you happen to know about that rule, and you're familiar with that, you can remember subtraction first, "h" first, and then vertical.

(Transcript, 9/11/89)

Shahroukh tried to clarify what he meant, but he seemed to realize he did not have the language he needed when he stated, "In some sort of math, you know, like I don't know what you call it, when you use brackets and have to use signs, division, those things?" His use of "I don't know what you call it" placed the burden for not knowing on him, not Lampert. Additionally, his comment, "I don't know what you call it, when you use brackets," indicated he thought Lampert probably did understand the concept he was trying to describe, as well as its mathematical term. In essence, he politely invited Lampert to provide him the specialized language needed to continue building his idea, and she did so. Once Shahroukh clarified his idea, Lampert signaled she understood it when she rephrased and summarized it for the class, "Okay, so one term that people use for that math is the order of operations. Subtraction, addition, division, and multiplication. So if you happen to know about that rule, and you're familiar with that, you can remember subtraction first, 'h' first, and then vertical."

Lampert's exchange with Shahroukh helped students begin to acquire their mathematical discourse in two ways. For listeners, Lampert modeled how to politely request information: First, assume the speaker understands his or her idea, and, second, acknowledge that you (the listener) don't understand and need more information. By phrasing her request to Shahroukh in this way, Lampert reduced the threat to his personal identity. The second way in which Lampert helped her students begin to acquire this new discourse was directed more toward future speakers. Her comments to Shahroukh let all students know that if a listener were to ask a speaker to clarify an idea, that listener was doing so in order to understand the speaker's idea in the same way the speaker did; asking for clarification was not to be taken as a personal challenge. Shahroukh's response to Lampert's request seemed to exemplify that idea as he did not appear defensive or nervous when he responded to her request; in fact, he in turn made a request of her. Lampert's talk highlighted the obligation both listeners and speakers had for trying to understand the conversation from the other's point of view. What mattered was that both listener and speaker needed to work together to build an understanding about mathematical ideas.

Scaffolding Students' Talk with One Another

Helping students develop polite interaction patterns for talking about mathematics was one way Lampert's talk fostered students' talk about mathematics. Helping students learn to talk with one another to build ideas – what Vygotsky (1978) calls scaffolding – was another.[7] Students sometimes have trouble talking with one another in classroom conversations, not because they can't (as listening to groups of children

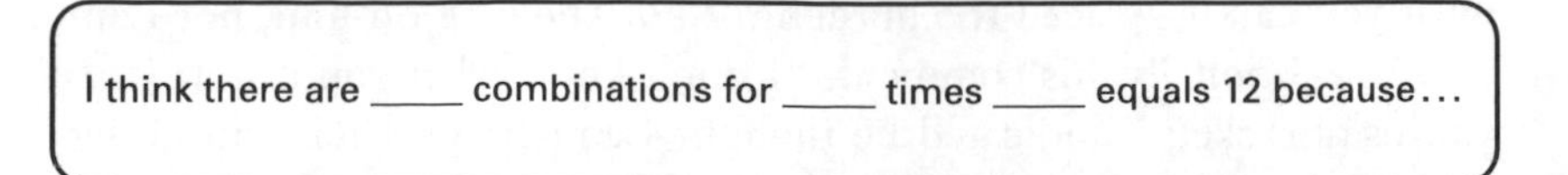

Figure 7.3. Problem of the day

outside of school demonstrates), but because in school they have been conditioned not to talk at all, or they have learned to route talk through the teacher. After all, the "default" discourse pattern (Cazden, 1988, p. 53) in most classrooms consists of the teacher asking a question, a student responding to that question, and the teacher evaluating that student's response before posing the next question (Mehan, 1979). When teachers ask students to move away from this usual pattern, students may not know how to do so. In this next segment, the students were in their large group discussing ideas about the problem of the day (POD) shown in Figure 7.3. As the segment begins, someone has just suggested that only numbers under 12 can be used to meet the conditions of the POD, and Lampert has asked her students to tell how they know that. She calls on Saundra.

Lampert: How do you know, some people are saying that you can only use numbers under twelve, how do you know the numbers you use have to be under twelve? Saundra?

Saundra: Because if you go over twelve and multiply it, I mean it's higher than, dividing, minusing, or adding to. When you times it, it'll go higher than twelve, then it'll get really high, so, it'll be over twelve.

(Transcript, 9/25/89)

Here Saundra tried to explain that multiplying "over twelve" would generate an answer that would be larger than if other operations ("dividing, minusing, or adding to") were used. Though unspoken, her comments indicate she believed multiplying by 12 would violate the problem parameters to find number combinations that equal 12. Thus, it would seem that only numbers under 12 could be used. At this point, Anthony joined the conversation, suggesting, "You don't have to always go below twelve." He seemed to disagree with Saundra, and at this point, Lampert stepped in to the conversation.

Lampert: Saundra what do you think of that? Anthony said, "You don't have to always go below twelve." Anthony why don't you turn around and say it again so Saundra can hear you.

Anthony: It's like, well, it's not the [inaudible] but if twelve times one it could work, so . . .
Saundra: Yeah, I know, but, I mean 'cause . . .
Anthony: Twelve times one is twelve.
Saundra: I mean, 'cause twelve times one is twelve, I mean everybody knows that, so I mean, like any higher than twelve, like twelve times something, then it would go over twelve.
Lampert: Okay, so . . .
Saundra: Like twelve times nine.

(Transcript, 9/25/89)

Lampert repeated Anthony's comment, then asked him to repeat it, "so Saundra can hear you." By placing so much emphasis on Anthony's comment, Lampert seemed to be signaling Saundra, and other students, to pay attention to this idea. Further, Lampert's suggestion that Anthony turn around so Saundra could hear him signaled both students that they should address their comments directly to one another. Saundra and Anthony did explore their ideas, working together to figure out what combinations might fit the conditions of the problem. However, their comments also indicate they were struggling with their ideas. Both appeared to stumble over themselves as they concurrently tried to think about their reasoning and explain it to others, and for good reason.

Saundra and Anthony were both engaged in exploratory talk (Barnes, 1990); they were busy trying to understand for themselves what numbers would fit the conditions of the problem. At the same time, they were also trying to explain their emerging thinking to one another. Both of these processes are hard enough on their own; that Saundra and Anthony were doing both concurrently added much complexity to their task. Even so, Saundra and Anthony did seem to be listening and thinking about what the other said. When Anthony restated his example, Saundra seemed to agree with him. When she rephrased her conjecture, she incorporated Anthony's condition. She still used words like "higher" and "over" to convey her sense that multiplying 12 by "something" could still violate the problem parameters. Working together, Saundra and Anthony refined Saundra's original conjecture to include the number 12 (but only under certain conditions – multiplying by 1).

Lampert's suggestion that Anthony address his comments to Saundra gave these students "permission" to talk directly to each other; they did not need to route their talk through her. More importantly, her comment nudged these two to join forces and see how the individual pieces of the puzzle each seemed to hold could be fitted together to clarify an idea.

Together, their comments seemed to help both Anthony and Saundra with the messy work of exploring an idea. Their talk was not polished, it was not presentational in nature, but it did help each of them think more deeply about the mathematical terrain of the problem.

Lampert's modeling of the kind of talk she wished her students to engage in and her use of scaffolding comments to help students begin to talk directly with each other helped make visible the kind of discourse the class needed in order to think and talk about mathematics more deeply. Through her modeling and nudging of students to talk with each other, Lampert helped immerse the class in more authentic talk about mathematics. For some students, this modeling and scaffolding helped make the conversation comprehensible enough that they could begin to engage in talk about mathematics. However, simply providing opportunities for more authentic talk did not guarantee students would acquire it. There were times when students needed information that only Lampert had, as in providing mathematical vocabulary or explaining how to argue. At times, Lampert had to step out of her role as participant and step into a role that more directly instructed students about the discourse they were learning. In the next section, I describe how Lampert did this.

Commenting on Mathematical Conversations: Talking About the Talk

As a commentator, Lampert helped her students think about the talk they used to do mathematics by "stepping out" of the conversation and talking about the talk. The kind of conversation Lampert wanted her students to engage in required them to analyze problems, make and support generalizations for those problems, and argue the merits of those generalizations with others. To do that, students needed to use mathematical terms and concepts, and as a commentator, Lampert provided students with this kind of information. Additionally, as a commentator, Lampert was able to stop the conversational action to draw her students' attention to what they were doing in order to help them understand how their talk was supporting their mathematics learning. In this section, I discuss the ways in which Lampert's talk helped students develop their abilities to explain their mathematical ideas and how to argue with one another about those ideas.

In the first section, I describe how Lampert's talk and efforts were directed at helping students tell why their conjectures fit the problem of the day. Asking students to engage in this kind of talk goes beyond the usual teacher question–student response–teacher evaluation model

Figure 7.4. Chalkboard, first problem of the day

of most classroom talk. Thus, Lampert's students likely were not accustomed to providing reasoning for their ideas and to get them to do so required Lampert to put this kind of talk "under the microscope." In the second part of this section, I examine how Lampert's talk helped students learn how to argue about the mathematical merits of their ideas. This, too, is not a common feature of classrooms. If students do venture their opinions, it is often in regard to something for which there is no "right" answer. Thus, all opinions are equally valid. The kind of justification Lampert asked students to provide went beyond personal opinion; rather, it was founded on the reasonableness and logic of the mathematics supporting the opinion. Students needed to challenge one another's ideas in order to work out the mathematical logic, but, again, since this is not often done in classrooms, they needed to learn how to do this.

Learning to Tell Why. On the first day of class, Lampert asked her students to consider the problem shown in Figure 7.4.

Lampert: 'Kay, let me have you look over here. These are two very easy additions and I bet everybody in the room knows the answer to this one and the answer to this one, but I don't want you to tell me the answer. I want you to tell me another addition problem whose answer is in between the answer to this one and the answer to this one, without saying the answers. What do you think? Can you tell me one? No answers, just an addition problem.

(Transcript, 9/11/89)

Specifically, Lampert asked her students to generate examples of addition problems that would fit between the two problems she had given them, and after some hesitation, they did so. The problems written within the circle area in Figure 7.5 are those the students nominated. Once the

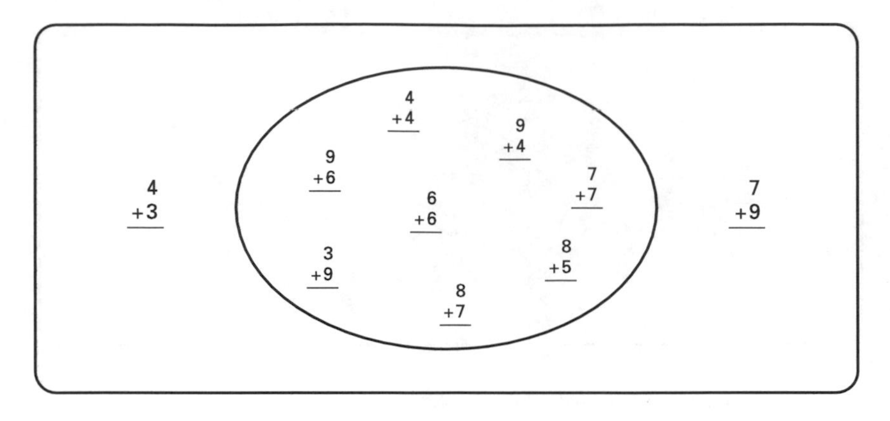

Figure 7.5. Chalkboard with student solutions

class generated their examples, Lampert asked them to explain why their examples fit the conditions of the problems. It was at this point that students began to have difficulty.

Lampert: Nine plus six. Now, I want you to look at these and see if you can tell me why all these work. How could you explain why all of these problems have answers in between these two? [Eddie and other students raise hands.] Eddie, do you want to try to explain that?

Eddie: No.

Lampert: No? At first you thought you would and then you changed your mind. Okay, Ellie?

(Transcript, 9/11/89)

When Lampert called on Eddie, he did not participate. Instead of asking him a series of probing questions to get him to speak, Lampert accepted his "No" and indicated this answer was an acceptable one. Her "At first you thought you would and then you changed your mind" seemed to tell Eddie (and the other students) that changing their minds about speaking was a perfectly normal thing to do. Further, by not evaluating his "No" as being impertinent or stubborn, Lampert also sent a message that "No" was an acceptable answer. In doing so, she seemed to be telling her students that she knew what she was asking them to do was different and that they might not be comfortable explaining themselves. Allowing them to "pass" helped create a sense that the environment she was trying to create for discussing ideas would be a supportive one.

In this regard, Lampert seemed to be creating a context that fostered risk taking (Dudley-Marling & Searle, 1991). She wanted students to feel they could put their ideas on the table without feeling personally attacked. Creating contexts where students feel safe to explore their ideas is important if we want them to talk seriously about those ideas and their merits. A context that supports risk taking, however, must do so without exposing students to a high degree of personal risk. Students can't play with big ideas if they feel personally threatened. Lampert seemed to understand that if she forced Eddie to respond, she would shut down just the kind of talk she wanted from him – and from the other students as well. So, she chose to accept Eddie's "No" and moved on to Ellie.

Ellie: Because, like, an example, four plus four that equals eight and eight is more than four plus three but it's less than seven plus nine.

Lampert: Okay, who can explain another one? Awad?

Awad: Um, because four plus three equals seven and you have to find a number that equals between those like, between seven and sixteen, so you're just supposed to [inaudible].

Lampert: Okay, so the answer has to be between the answer to this one and the answer to this one. Can anybody else pick one of these out and explain to me why it's true? Shahroukh?

(Transcript, 9/11/89)

As Lampert elicited comments from Ellie and Awad, she continued to accept what they said, even though their comments weren't really explanations. For instance, Ellie gave an example; she did not explain a general mathematical rationale for *why* her example worked. Awad, too, did not really provide an explanation; rather, he simply restated the problem. Lampert accepted these comments, however, and in so doing continued to convey to the class that there existed a low level of personal risk for those who chose to speak. However, Lampert also began to nudge students who seemed reluctant to "give it a try."

Shahroukh: Eight plus seven is fifteen and it's between four plus three and seven plus nine.

Lampert: Okay, Ivan, can you explain another one? Want to try?

Ivan: He took mine.

Lampert: I bet you can explain this one. Can you? Don't think so? [Yasu], do you want to try to explain this one?

Yasu: Nine and four is thirteen and thirteen is between the answer of these two problems.

Lampert: That's a good explanation. Thirteen is between the answer to these two problems.

(Transcript, 9/11/89)

Here, Lampert called on Ivan, but when he tried to pass by saying, "He took mine," she stayed with him, pointing to another problem on the board and saying, "I bet you can explain this one. Can you? Don't think so?" Even with her repeated nudges, Ivan still seemed reluctant, so Lampert moved on. Her comments here seemed to tell students to try their hand at taking risks. Ivan's response and unwillingness to try indicate that, for some students at least, giving explanations was still just too risky, and Lampert backed off. For those who did try, however, Lampert provided supportive feedback through her "Okay," which seemed to take on more of a "keep going" function rather than an evaluative one, and her only evaluative comment, "That's a good explanation," was a positive one. By not criticizing what students had to say, Lampert let students know it was safe to talk, as the next segment indicates. After Lampert's comment to Yasu (above), she called on a student whose hand was raised. That student was Eddie, who earlier said "No" when Lampert called on him.

Eddie: I think there's any number below nine times seven it would be lower than that nine times seven.

(Transcript, 9/11/89)

Eddie's statement is important because it is the first real attempt at providing an explanation for why any problem that worked did so. Would Eddie have ventured a comment if Lampert had earlier insisted he answer her? We'll never know. It seems likely, however, that he did listen to what Lampert said (and didn't say) and to what she did (and didn't do) as she elicited comments from the class and worked to get the students to explain their examples. That Lampert spent this kind of time and effort helping her students feel comfortable taking risks in telling why they believed their ideas were valid may have sent the kind of signal Eddie needed to join in the conversation. Further, by not calling on him earlier, Lampert likely provided Eddie the time and space he needed to think about this request of Lampert's. The result was that Eddie's comment began to move the class conversation in a different direction, away from example citing and toward theory building.

Learning to Argue. As soon as Eddie stated, "I think there's any number below nine times seven it would be lower than that nine times seven," Anthony immediately chimed in with, "times seven?" While he might simply have been trying to help, Anthony's comment was not unlike those of students who seem more interested in showing up their peers than in thinking about the ideas under construction. This game of "gotcha" could have shut down the conversation Lampert was trying to foster and Lampert immediately froze the conversational frame in order to help her students learn about the nature of mathematical disagreement.

Lampert: Okay. Now here we have a very interesting exchange, which is something I think is going to happen and hope happens a lot in this class. But we need to have some ways of having this exchange. Eddie said, "I think anything lower than nine times seven is going to have an answer that works." And Anthony said, without raising his hand, "times seven?" Okay, and then Eddie revised and he said, "plus seven." If you disagree, like Anthony just disagreed with Eddie, that's very, very important to do in math class. But, when you disagree or think somebody misspoke, you need to raise your hand and say, "I think he must have meant plus not times." And then Eddie will probably revise even before you get it out of your mouth. So one thing we have here is how to challenge or disagree with somebody in your class, and that's a very important thing. Mathematicians do it all the time, but you have to have a good reason and you have to do it with politeness. But the other thing I want to come around to is really what Eddie meant. Eddie, why don't you say again what you're trying to say here.

(Transcript, 9/11/89)

Lampert's comments revealed to the class two important issues she wanted them to understand about the kind of talk they were learning. First, Lampert told the class she wanted students to challenge each other's thinking and ideas, as occurred between Eddie and Anthony, "Okay, now here we have a very interesting exchange, which is something I think is going to happen and hope happens a lot in this class." Her comment let the class know she placed a high value on this kind of disagreement. She said Eddie's and Anthony's comments represented a "*very* interesting exchange [emphasis added]" and one that she hoped would happen "a lot in this class." She again emphasized the value of the exchange when she said disagreeing was "very, very important to do in math class," and that "Mathematicians do it all the time."

Lampert's comments provided an opportunity for her students to begin to think differently about disagreement. In many classrooms, teachers try hard to help their students learn how to get along with one another. Learning to "get along," however, ought not to mean seeing eye to eye on everything. Life isn't that way. Part of getting along with others means learning how to respect others' ideas and talk civilly about those ideas. This is true in the intellectual life of disciplines as well, but not simply because it's nice to get along with everyone in your field. Disagreement is an essential component of constructing disciplinary knowledge. In stopping the conversation and pointing out the importance of disagreement, Lampert was letting her students know that disagreement is a normal, natural, and essential part of mathematical talk. However,

Lampert also let the class know that disagreement for its own sake wasn't acceptable: "You have to have a good reason [to disagree], and you have to do it with politeness." The issue of politeness was the second important point Lampert made about disagreement.

Lampert's comments provided her students with information about how to disagree. First, she let them know there was a procedure to follow, "When you disagree or think somebody misspoke, you need to raise your hand and say . . ." Reminding students to raise their hands to speak implies Lampert wanted the class to route their talk through her, and this seems to contradict her earlier efforts to have students talk directly with one another. Having in place a procedure that slowed down students' reactions to one another's statements, however, provided a way of maintaining civility among students. A student who had to be acknowledged by Lampert in order to disagree with another student would be accountable to Lampert; she would know who said what. Thus, raising a hand likely would cause students to think twice about what they said and how they said it. Further, raising hands helped reduce the chances of a group of students verbally tackling someone with whom they disagreed. Thus, hand raising served to help preserve a speaker's self-esteem when confronted about his or her ideas and fostered a sense of responsibility on the part of those wishing to disagree. This idea was elaborated on when Lampert told the class that when they did disagree, they had to "do it with politeness." She made it clear that challenging one another's ideas was not an open invitation to attack someone. Once Lampert clarified the norms for disagreeing, she asked Eddie to repeat his statement before commenting on it once again.

Eddie: Okay. If there is any number below seven plus nine like an addition problem that has [inaudible] seven plus nine and [inaudible] but it has, like, the numbers have to be, the numbers, well they could be lower than four plus three and sometimes they have to be higher.

Lampert: Okay, so what Eddie is trying to do here is make a generalization about what will always work, and he put a lot of conditions on his statement. Let's hear from some other people. Awad?

(Transcript, 9/11/89)

Lampert asked Eddie to restate his idea so that she could name what he was doing, "Okay, so what Eddie is trying to do here is *make a generalization* about what will always work [emphasis added]," and describe the statement's features, "and he *put a lot of conditions on his statement* [emphasis added]." Holding up Eddie's statement for other students to see showed the class that that kind of talk was something Lampert valued. Further, by naming the talk, she identified it as a special kind of talk, something that helped make mathematical talk "mathematical" in nature. When Lampert stepped out of the conversation to name its components (generalizations), make clear its essential qualities ("If you disagree . . ."), and clarify its norms (". . . you have to do it with politeness"), her talk helped make the discourse more comprehensible for students.

Summary

Lampert's talk and her dual roles as participant and commentator provide us insight into one teacher's vision of what fostering students' understanding of mathematics might look like. Clearly, this kind of teaching is not easy, for it asks teachers to move away from a more traditional role of teacher as teller toward a role of teacher as facilitator. This kind of teaching also asks teachers to view mathematics as more than the memorizing of algorithms. It asks teachers to view mathematics and mathematics teaching as opportunities to engage students in problem posing and problem solving; as opportunities to reason and conjecture about the relationships between numbers. Creating contexts where students work and talk together to solve problems has the potential to help all students engage in the kind of cognitive apprenticeship that more closely links factual and conceptual knowledge to actual use. However, providing students opportunities to more fully participate in this kind of mathematics learning depends on teachers recognizing the importance of mathematical talk. Without it, students' opportunities to develop mathematical power will be diminished.

Acknowledgments

The research reported in this chapter was supported by the Spencer Foundation under grant no. 1992–00088 with additional support from the National Science Foundation under grant no. TPE–8954724. The opinions expressed do not necessarily reflect the views of either foundation.

Notes

1. For a discussion of what has tended to occur in mathematics classrooms, see NCTM (1991) and Goodlad (1984).
2. For a discussion of how tasks and routines are used, see Rittenhouse (1997).
3. Allwright and Bailey (1991). See their illustration of input with regard to learning a new language for a more complete discussion of this idea.

4. See, for example, Lave and Wenger (1991); Rogoff (1990); Collins, Brown, and Newman (1989).
5. Cited in Rogoff (1990), p. 39.
6. For a discussion of managing equally valid yet conflicting goals, see Lampert (1985).
7. Scaffolding is something that falls between Lampert's two roles of "participating" and "commenting." Embedded in it are elements of both kinds of teacher talk that are so entwined as to be impossible to separate. I have chosen to discuss scaffolding under participation because it did not seem to stop the conversational action in ways that Lampert's commentator comments did. Rather, from my perspective, Lampert's efforts to scaffold student talk seemed woven into the conversational flow. I provide this bit of background information so that others will better understand my thinking about this.

References

Allwright, D., & Bailey, K. M. (1991). *Focus on the language classroom: An introduction to classroom research for language teachers.* New York: Cambridge University Press.

American Heritage Dictionary (1983). (2nd College ed.). Boston: Houghton Mifflin.

Barnes, D. (1990). Oral language and learning. In S. Hynds & D. Rubin (Eds.), *Perspectives on talk and learning,* pp. 41–54. Urbana, IL: National Council of Teachers of English.

Brown, S. I., Cooney, T. J., & Jones, D. (1990). Mathematics teacher education. In W. R. Houston (Ed.), *Handbook of research on teacher education,* pp. 639–656. New York: Macmillan.

Cazden, C. B. (1983). Adult assistance to language development: Scaffolds, models, and direct instruction. In R. Parker & F. Davis (Eds.), *Developing literacy: Young children's use of language,* pp. 3–18. Newark, DE: International Reading Association.

Cazden, C. B. (1988). *Classroom discourse: The language of teaching and learning.* Portsmouth, NH: Heinemann.

Collins, A., Brown, J. S., & Newman, S. E. (1989). Cognitive apprenticeship: Teaching the crafts of reading, writing, and mathematics. In L. B. Resnick (Ed.), *Knowing, learning, and instruction: Essays in honor of Robert Glaser,* pp. 453–494. Hillsdale, NJ: Erlbaum.

Dudley-Marling, C., & Searle, D. (1991). *When students have time to talk: Creating contexts for learning language.* Portsmouth, NH: Heinemann.

Gee, J. (1991). What is literacy? In C. Mitchell & K. Weiler (Eds.), *Rewriting literacy: Culture and the discourse of the other,* pp. 3–11. New York: Bergin & Garvey.

Goodlad, J. I. (1984). *A place called school: Prospects for the future.* New York: McGraw-Hill.

Kitcher, P. (1986). Mathematical change and scientific change. In T. Tymoczko (Ed.), *New directions in the philosophy of mathematics,* pp. 215–242. Boston: Birkhauser.

Krashen, S. (1977). Some issues relating to the monitor model. In H. Brown, C. Yorio, & R. Crymes (Eds.), *On TESOL '77, teaching and learning English as a second language: Trends in research and practice,* pp. 144–158. Washington, DC: TESOL.

Krashen, S. (1982). *Principles and practice in second language acquisition.* Oxford: Pergamon Press.

Krashen, S. (1985). *The input hypothesis: Issues and implications.* London: Longman.

Lakatos, I. (1976). *Proofs and refutations.* New York: Cambridge University Press.

Lampert, M. (1985). How do teachers manage to teach? *Harvard Educational Review, 55* (2), 178–194.

Lave, J. (1988, May). *The culture of acquisition and the practice of understanding* (Report no. IRL 88–0007). Palo Alto, CA: Institute for Research on Learning.

Lave, J., & Wenger, E. (1991). *Situated learning: Legitimate peripheral participation.* New York: Cambridge University Press.

McLeod, A. (1986). Critical literacy: Taking control of our own lives. *Language Arts, 63* (1), 37–49.

Mehan, H. (1979). "What time is it, Denise?": Asking known information questions in classroom discourse. *Theory into Practice, 18* (4), 285–294.

National Council of Teachers of Mathematics (1989). *Curriculum evaluation standards for school mathematics.* Reston, VA: National Council of Teachers of Mathematics.

National Council of Teachers of Mathematics (1991). *Professional standards for teaching mathematics.* Reston, VA: National Council of Teachers of Mathematics.

Rittenhouse, P. S. (1997). *Orienteering across the mathematical landscape: Using tasks and routines to foster student talk about mathematics.* Unpublished manuscript.

Rogoff, B. (1990). *Apprenticeship in thinking.* New York: Oxford University Press.

Scribner, S. (1984). Literacy in three metaphors. *American Journal of Education, 95,* 6–21.

Stedman, L. C., & Kaestle, C. F. (1991). Literacy and reading performance in the United States from 1880 to the present. In C. F. Kaestle, H. Damon-Moore, L. C. Stedman, K. Tinsley, & W. V. Trollinger, Jr. (Eds.), *Literacy in the United States,* pp. 75–128. New Haven, CT: Yale University Press.

Venezky, R. (1991). The development of literacy in the industrialized nations of the west. In R. Barr, M. L. Kamil, P. Mosenthal, & P. D. Pearson (Eds.), *Handbook of reading research,* pp. 46–67. White Plains, NY: Longman.

Vygotsky, L. (1978). *Mind in society: The development of higher psychological processes.* M. Cole, V. John-Steiner, S. Scribner, & E. Souberman (Eds. and Trans.). Cambridge, MA: Harvard University Press.

8 Teacher Talk About How to Talk in Small Groups

Merrie L. Blunk

The recent trend in mathematics educational reform includes an emphasis on small groups. Lindquist (1989), for example, argues that the use of small groups in mathematics teaching and learning can encourage verbalization, increase students' responsibility for their own learning, encourage students to work together to build social skills, and increase the possibility of students solving certain problems or looking at problems in a variety of ways. Indeed, some researchers propose that it is only through communication and participation in a community that novices learn what it means to be a member of that community (Lave & Wenger, 1991).

While a great deal of research has been done investigating the learning that occurs within small groups and the various factors that influence this learning (e.g., Palincsar & Brown, 1984, 1986; Palincsar, 1986; Yackel, Cobb, & Wood, 1991; Russell, Mills, & Reiff-Musgrove, 1990; Johnson, Johnson, & Skon, 1979; Webb, 1983; Cobb, 1995, to name but a few), there have been few studies which have investigated the teacher's role in small groups. Studies that describe small group formats used by teachers and that identify effective small group instructional processes are practically nonexistent (Good, Grouws, Mason, Slavings, & Cramer, 1990). One such area that deserves further attention is the teacher's role in creating and maintaining effective small groups.

Can fifth graders learn to work productively with peers without the direct supervision of the teacher? The M.A.T.H. Project has videotapes of most school days in one classroom during the 1989–90 school year. On the majority of these days, the children were asked to work in small groups, and from the videotapes, it is apparent that the students do talk mathematics productively with each other. For example, on one day a group of students were talking about two conjectures fellow students had made the previous day: Ellie's conjecture was $.50 + .50 = 100$ and Sam's conjecture was $.5 + .5 = .100$.[1]

Candice: I disagree [about Sam's conjecture] because if there's one number after the decimal, you have to round it to its nearest, like I don't know how to explain it. Like it goes to ten. Like point five plus point five equals point ten.

Shahroukh: I don't think so.

Candice: It would. Why not?

Shahroukh: Point five plus point five . . . point five is one half, and one half plus one half equals one whole, isn't it?

Candice (& ?): And point ten is a whole.

Shahroukh: I think point ten is one tenth of a whole.

Candice: Point ten is one whole, point one zero, zero is one whole, point one zero, zero, zero is one whole. If there are three numbers after a decimal, the number is going to be out of, 'kay, let's say if there are three numbers after the decimal, right. Let's say something is point nine, nine, five or something, 'kay? It would be like, out of a thousand, out of a thousand pieces.

Donna Ruth: Yeah, and that would be one whole.

Candice: You would have however many pieces, three numbers after the decimal of pieces out of a thousand pieces. If there is one number after the decimal, the cake is cut into ten pieces. If it's two numbers, the cake is cut into one hundred pieces.

Shahroukh: I disagree.

Candice: Why?

Shahroukh: I'll explain. First of all, when you said like point nine, nine, five will have to be under a thousand? Like do you mean . . .?

Candice: Out of a thousand. Point nine, nine, five out of point one hundred, no point one thousand.

Shahroukh: Let's see if I get what you were saying. Point nine, nine, five will go under one thousand.

Candice: Point one thousand.

Shahroukh: Point one thousand. Hmmm.

Candice: Uh, huh. Because this can't be under one hundred, because it's over one hundred. Well, it could be.

Donna Ruth: Yeah. It's way over one hundred.

Shahroukh: Okay. In decimals, this is ones. No, this is the tens . . .

Candice: It could be hundreds.

Shahroukh: This is the hundreds in this case, okay.

Candice: Uh-huh.

Shahroukh: This is the tens, and this is the ones. Okay? So what this is saying . . ., let's say it's ninety-five cents, okay. Ninety-nine cents. If I had ninety-nine cents, that wouldn't be out of a hundred, would it?

(At this point the students talk about the relevance of using money to represent decimals, but then abandon the comparison.)

Candice: So what you [Shahroukh] are saying if there are three numbers after a decimal . . .

Shahroukh: It doesn't matter how many there are.

Candice: I know!

Shahroukh: I could have a number that is called point nine, nine, seven, five, two, one, zero, one, okay? That will be . . . that's still less than one.

Candice & Giyoo: Right. I know.

Donna Ruth: One is larger than this number?

Giyoo: It is because it's nine, and nine is not one whole. Point nine is smaller than one whole.

Shahroukh: You get it? Yeah, you get it.

Candice: 'Cause ninety-nine, well no . . .

Shahroukh: One is larger than all this.

All: Yeah, yeah.

Shahroukh: One is larger than anything. Like if this is all nine . . .

Candice: It's just one thing larger (similar comments from rest of group).

Shahroukh: One point zero, zero, zero, zero, zero, zero (group chimes in on zeros) is larger. If you had point nine, nine, nine all this, plus point zero, zero, zero, one, you will get exactly one. Right.

Giyoo: So what do you say about Sam's conjecture?

Shahroukh: Well I disagree. Point five, zero.

Candice: Sam said point five.

Shahroukh: It doesn't matter if it is point five or it doesn't matter if it is point five, zero. It doesn't matter.

Giyoo: Shahroukh, one hundred. Shahroukh. Shahroukh. Shahroukh.

Candice: I still don't understand.

Shahroukh: Okay. Sam said point five, right. Okay. Plus point five, right. Let me show you how to add this, it's important.

Candice: That would be point one zero. It would be one whole.

Shahroukh: Five plus five is . . .

Candice: Ten. One point zero, right?

Shahroukh: So this would be ten, right? See, pretend that this thing [decimal point] is like dropping down, this is dropping, dropping. Will it drop right here?

Candice: Yeah.

Shahroukh: One will go right here.

Candice: That's what I said. One point zero.

Shahroukh: Right, so it can't be point one, zero, the decimal will always line up, down.

Reba: So, he says that it's a whole.

Shahroukh: You can't . . . Look at Ellie's problem. Look. Point five, zero plus point five, zero, zero, zero, point falls down, alright.

Candice & ?: Yeah.

Candice: So it's one.
Shahroukh: Right. See, one point zero, zero.
Donna Ruth: Oh.
Shahroukh: The decimal will not change its place.
Candice: The decimal cannot go diagonal. That's what he is saying. The decimal cannot go down on a curve.
Giyoo: So, what you are saying is that it doesn't equal point one, zero, zero, right?
Shahroukh: No. It equals one point zero, zero.
Giyoo: Ah!!! So you are saying that for both problems. So you are saying it equals a whole.
Donna Ruth: Oh!!! Okay!!
Shahroukh: Right. One whole.
Candice: In Sam's if it, . . . one point zero and in Ellie's one point zero, zero.
Shahroukh: One point zero and one point zero, zero means still one.
Candice: Right. One with nothing left.

This is a complex and intricate conversation for such young students. What characteristics of this classroom create an environment in which children are able to carry on such a sophisticated conversation with no direct intervention from the teacher? After all, in most classrooms, this type of discussion among children is rare (Gerleman, 1987; Good, Grouws, & Mason, 1990). One way to investigate this question, the one used here, is to look at what the teacher says to the students about small groups. This approach is chosen because it presents a picture of what the students themselves might experience being in this classroom. It is assumed here that from the students' point of view, a working understanding of what is in the teacher's head, and therefore what is expected of them, can come from what the teacher says to the students. If she tells them that she will grade their work based on quantity, this implies they should take a different approach to a task than if she says she will grade based on quality of work. In other words, what the teacher actually says to the students is one of the factors affecting what the students do within the classroom. If this is the case, then we can assume that the behavior of the students within small groups is determined in part by what the teacher teaches about small groups.

Because there exist videotapes of almost every day Lampert taught during the 1989–90 school year, it is possible to have access to what Lampert taught to the students about small groups. The lesson transcripts provide an opportunity to observe not only how this teacher establishes small groups early in the year, but also how she maintains them. Since so few data sets have been available across the school year, this study has a unique strength which other studies looking at small groups have thus far not had.

Method

My initial entry into the M.A.T.H. Project data was with a study group of teachers, teacher educators, and graduate students who met on a weekly basis to talk about issues related to teaching and learning in the data. Through this group, I developed an overall view of the curriculum, instruction, and students within the Lampert classroom. This focused study followed from the perspective of the classroom that I developed from this group.

In order to capture what the teacher said to the students about small groups, I searched all the lesson transcripts using a keyword method, using the Data Calendar tool created by a M.A.T.H. Project programmer. I searched transcripts for particular keywords related to small groups. My first search involved finding instances of the term *small group* in the lesson transcripts. While I found several instances of the phrase, it was apparent that Lampert used other terms when talking to the class about small groups. After reading many of the transcripts and searching on various terms, the following list of key terms was generated: *small group, group, table, pairs,* and *partners.* All instances of the words on the list were identified in the transcripts and collected into a smaller data set of instances in which Lampert made a statement to the class about small groups. This data set captures the vast majority of instances in which Lampert referred to small groups in the lessons. What is missing from this data, however, are the instances in which Lampert referred to particular groups or pairs of students by name. Because of the difficulty in searching all the lesson transcripts for particular students' names when used only in reference to small group work, these instances are not well represented in this data.

From searching all the lesson transcripts, an extensive list of teacher statements was generated and collected into a "book of quotes." Sometimes these statements or quotes were just a sentence, in other cases they were several sentences or a paragraph. Whenever the quote included different statements about small groups, the statements were considered individually. After the quotes were gathered together, several categories became apparent: types of statements Lampert made about small groups. Early in the semester, Lampert's teaching about small groups was heavily concentrated on what small groups are and why she was asking students to work in small groups. After this introduction and description of small groups, Lampert's statements to the class about small groups became more focused on how she wanted students to work while in their groups. When describing how she expected students to work, Lampert talked about what she considered the rules of behavior in working in groups, how to cognitively work together on a task, how to socially work together, and how she would evaluate how well the students were working together.

Each of the categories described above – why small groups were being used in this classroom, what and who small groups were, and how the teacher wanted the students to work in their small groups – will be discussed in turn. This organization provides a somewhat temporal sequencing of Lampert's teaching about small groups to her class; however, some particular quotes or comments are not in strict chronological order. Some instances are indicated in which the timing of Lampert's statement seems particularly important.

Why Work in Small Groups?

On the day that Lampert first introduced the idea of small groups to the class in mid-September, she described her reasons for having students work together in small groups:

> From now on, when we're having individual work, when you have a problem, like if you don't understand what you're supposed to be doing or you wanted to discuss a conjecture with somebody, you should ask the people in your group. What that means is you don't sit there and don't do nothing, waiting for the teacher to come along and answer your question. You ask the people in your group and that's very important. Now sometimes people are a little bit shy about asking other classmates for help. 'Cause you think oh, you know, if I ask that person for help they'll think I'm dumb or something like that, but that's part of working in a human group. The people need to get used to asking each other for help when they need it. They need to get used to asking questions. School time is very, very precious and teachers can't get around to helping everybody and there are a lot of people in this class who can help besides the teacher and so I want you to get used to the idea of asking for help and giving help to the people in your group. (Lesson transcript, 09/20/89)

What might the teacher have communicated to the students about this class and herself based on this statement to the class? First, the teacher implied that she respects her students since she is willing to be so explicit with them in discussing her reasons for students to work in groups. Second, the students heard that this teacher believes that peers are important resources not only because class time is limited, but also because students need to practice developing the skills of working with others. Third, the teacher made it clear that she did not believe that she was the sole originator of knowledge in the classroom. Acknowledging students' own understanding and sense-making abilities, she encouraged students to ask each other for help with directions as well as with problem solution. The implication, again, is that one can learn from peers as well as from the teacher, and given the teacher's limited time, one is more likely to receive help by talking with another student instead of just waiting with a hand up for the teacher to come by and answer the question. Finally, the teacher's early statement to the class expressed sensitivity to potential student discomforts or difficulties in asking other students for help.

Throughout the school year, the teacher made additional comments to the class about her reasons for having students work in small groups. In one instance, she told the students that part of learning is becoming less dependent on the teacher, that students should get encouragement and information from the other people in their group instead of waiting for the teacher. She even specifically told one group that they should try to work independently of her as the other groups were. This is an unusual approach to teaching – handing responsibility over to the students for their own learning and understanding and in fact encouraging them to work independently of the teacher.

What and Who Are Small Groups?

In addition to talking to the students about why she wanted them to work in small groups, the teacher also talked about what she meant by "small groups." Some of her comments described the nature and characteristics of small groups through which she defined small groups in her classroom. Other times she defined small groups by their composition; the small groups were collections of particular students assigned by the teacher to work together.

Characteristics/Nature of Small Groups. On several occasions over the year, Lampert shared her ideas with the class about what small groups were to be like in this classroom. The vast majority of these comments were made early in the year.

On the first day of class as the regular classroom teacher, Thom Dye, along with Lampert, shared his expectations about classroom culture, Dye told the class: "When we're in the classroom and it's formal or we have guests come into the room . . . I would like you to address me in a formal way, that would be Mr. Dye. But when we're working . . . in small groups together, we're trying to share ideas so that's informal so you can call me Thom there, if that is comfortable for you" (9/7/89). His comment set the tone for how students were to interact while in small groups – that the formal roles of teacher and student were relaxed and the kind of communication that went on could be different from what happens when the class is operating as a whole. Lampert also included herself as both student and teacher during small group time by adding to Dye's comments: "When we are working in small groups we share ideas with each other." Shortly thereafter, Lampert also told the class that it was important to ask people in your group for help since she could not get around to everyone during small group time. Together these comments describe working in small groups as a productive, sharing time, during which there are no distinctions between participants. This is an unusual approach to small group work (Good, Grouws, & Mason, 1990). Often the teacher is viewed as an authority by small groups; when present, he or she is there to answer questions, not to share ideas. In addition, small groups are often created with the purpose of completing some task at hand – for the students to complete worksheets or quiz each other for a spelling test. In most classrooms this type of small group work is seen as less valuable than work done when the teacher is present. Individual time from the teacher is highly valued and desired. In this classroom, however, the teachers told the class that not only would they not always be in the groups but students' talking and listening to other students was a valuable and desirable source of learning.

Indeed, not only did Lampert describe the time spent working with others as valuable, but she said that such work is even more valuable than getting correct answers. Lampert specifically told the class that "it is okay if you need to copy something out of someone's notebook." This is a fairly unprecedented idea within classrooms – that students be allowed to copy each other's work. That Lampert not only allowed this practice but actually encouraged it illustrates that she was more concerned with students' reasoning and explaining than with their getting correct answers.

Another characteristic Lampert used to define small groups involved mutual responsibility among group members to understand and to be understood. For example, on one day Lampert told the class that everyone needed to work together in the small groups:

When you try to make an explanation and you try to get people to understand, everybody has to work on it together. The person who is explaining has to try to make the other people understand. The people who don't understand have to ask questions. So if the person who is explaining says a word and you don't know the meaning of that, you need to say, "But what do you mean by that?" (Lesson transcript, 11/9/89 B)

Lampert also talked to the students about the development of the groups. Her statements made it clear that not only do students need to learn how to work with other students but the small groups, as entities in themselves, take time to develop. As other research has shown (e.g. Blunk, 1996; Kantor, Elgas, & Fernie, 1993), there is a developmental nature to small group work. In early November soon after new small groups had been created, Lampert told the class that during small group time earlier in the class period she had told one student to explain his answer to only one other student in the group instead of the whole group because that group was just beginning to figure out how to work together. She seemed to be telling the class that learning to share ideas and explanations with several other people takes practice and is not something that students are assumed to be able to do well the first time they try.

On one day in March, Lampert summarized her ideas about the nature and characteristics of groups and her role in them. She told the class: "When you talk in small group discussion, what you need to worry about is whether people can understand what you are saying, and it's my job to help you learn how to get other people to understand what you are saying." The teacher viewed her role in the classroom as being a facilitator of social and cognitive skills of communication about mathematics, not simply a giver of knowledge and information (see Rittenhouse, chapter 7). In addition to these roles, the teacher had the role of assigning student groups.

Assigning Students into Small Groups. Lampert was responsible for organizing students into small groups for math. At times the students were asked to work with one other person, who the teacher defined as a partner, and other times students were asked to work together as a group, which was composed of the four to six students at a table. Students who happened to end up sitting by themselves, either because they had been moved by Lampert for management purposes or because their partners or others at their table were absent, were always reassigned seats during math lessons so that no student was denied the opportunity of having someone with whom to work.

Lampert was aware of the difficulties in having students work together. Unlike many teachers, however, Lampert was unusual in her approach of making explicit to the students her thoughts about how and why she put particular students together as partners or groups. At one point, Lampert and Dye spoke to the class about their reasoning for retaining control over assigning students to groups. On this occasion, Dye told the class that he and Lampert were thinking about changing some seats later in the month but that he was happy with the current arrangements, that they were all working so well together. However, he said, "We want all of you to work with different people so that you get used to communicating with different people." Both teachers were concerned with having students learn to work together, that as young mathematicians they begin developing their skills in explaining themselves and working to understand what other people are attempting to explain. Although it is clear that Lampert wanted small group time to be effective in terms of learning and struggling with mathematics, she also made it clear that she felt it was important for students to have a variety of experiences in dealing with different kinds of people so they could develop their communication skills.

How to Work in Small Groups

Throughout the school year, Lampert often spoke to the class about her expectations for how the students were to work together. There were more of these kinds of comments than any other about small groups. Early in the school year when the students were first beginning to work in small groups, Lampert introduced the class to a particular set of expectations that she defined as "the small group rules." In addition to these formal rules, the teacher described other expectations to the students during the school year regarding both academic and social aspects of working in groups. She also shared her ideas about how students should work together by evaluative comments directed to particular students or groups of students as they went about their work together. Each of these categories of comments will be described separately below, although there is overlap between them. For example, there were many instances in which the teacher talked about both cognitive and social aspects of working together at the same time.

Rules for Working in Small Groups. The first time Lampert talked to the students about how to work together in small groups occurred two weeks into the school year. Up to this point the students had not worked together in small groups. Lampert spoke about why it was important for students to learn to talk to each other and about how she expected students to talk. She also described what she called the small groups rules.

The rules are 1) that students are responsible for their own behavior, 2) that they be willing to help anyone in their group who asks, and 3) that they may not ask the teacher for help unless everyone in the group has the same question. For most students, these rules are quite different from the rules they have had in other classes, rules that are more likely to be about not talking with other people while you work and how, when you have a question, you should sit quietly with your hand raised until the teacher can come to you. According to Lampert's rules, students are expected to use their peers as resources for help and information and, in turn, be willing to help other group members; in fact, peers are to be the first place to go for help and are not to be bypassed to get to the teacher. In other words, group members are students' primary means of help, not the teacher.

One important aspect of these rules is that they reflect social as well as academic expectations about small group work; neither is promoted at the expense of the other. This dual set of concerns remained consistent throughout the school year, though at different times there appeared to be more emphasis on one than on the other.

There were variations that occurred in the small group rules throughout the school year, but the three components remained in effect through the end of the year. In fact, Lampert reminded the class of the rules, particularly the rule about asking each other for help before asking the teacher, at least once a month during the year, including June. Sometimes Lampert did this by restating the rule and other times she did it by asking a particular student who had requested her help whether or not he had already asked other people in his group for help.

Another component of classroom structure that Lampert used in small groups was her requirement that the class keep notebooks of their work for the year. Each day, the students were to write down the problem of the day, along with their answers and their reasoning, and any experiments or extra problems they had created to work on when they were done with the problem. Thus, when students were done writing down their reasoning about the problem, they had the option of offering their help to others in the group or creating extra problems or experiments. Several times over the year Lampert reminded students that they were not done with their work when they finished the problem, but that it was their responsibility to help other people in their group who might be having difficulties. In fact, Lampert on one occasion told one group of students who had finished a problem and closed their notebooks that when they were done with a problem, they were to leave their notebooks open in case someone else wanted to see how they had done the problem.

Lampert redefined the kinds of behaviors that she required of the students as they worked together throughout the year. For example, on one day, Lampert told the class that although the general rule was to stay in one's own group, it was okay to do some communicating across groups. Lampert's willingness to revise the small group rules and allow for some exceptions illustrates her responsiveness to student needs, both academic and social. These two aspects of communication within small groups, cognitive or academic and social, were clearly defined and described by Lampert throughout the school year. As discussed under the section below on how to work in small groups, Lampert carefully set the groundwork for effective interactive communication among students in their small groups.

By November, when the students were doing more cognitively difficult tasks within their small groups, Lampert added an additional expectation: If everyone in the group agrees on a solution to a problem, each person has to be able to explain how he or she got the answer. This statement illustrates Lampert's assessment in context – providing increased autonomy based on the judgment that the students are able to handle increased responsibility. Lampert's statement also illustrates one of her apparent philosophies – that the important thing in math is not having the right answer but rather being able to think through a problem and explain how one reached a solution.

One revision to Lampert's small group expectations came as late as early June. The students were actively engaged in a problem about tangrams and were comparing the number of shapes each group could find. Students were becoming competitive in their comparisons, at which point Lampert told the class that the goal of the day's activities was to learn about shapes and their relationships, not to be aggressively competitive between groups. She went on to say that not only was this competition not what she expected, but that she would rearrange things to stop the competition should it continue.

The value of Lampert's decision to spend time with students teaching them how to work with one another is supported by research. There is evidence, for example, that groups that have been trained in small group interaction engage in more task-related interaction than nontrained groups (Swing & Peterson, 1982). Work by Palincsar (e.g., Palincsar & Brown, 1984, 1986; Brown & Palincsar, 1982) also indicates that this is indeed the case. In her work, students are taught how to interact in ways that encourage participation and learning from all members. Her results are consistently positive: In relatively short periods of time, students come to resemble teachers in their behavior with each other in the group.

How to Work with a Partner: Academic Task. The largest category of comments Lampert made to the class about small groups was her descriptions of how she wanted students to work together in terms of the academic tasks and their cognitive requirements. These comments ran consistently from early September through the last day of class in June. Often the description of how to work together was specific to a given task. On the first day of group work, for example, Lampert told the students that she wanted them to see how many examples each table could come up with of a certain type of problem and to determine whether the problems each student had were the same as the problems the person sitting next to them had. At other times the comments were broader and referred to general expectations of how students were to interact with each other while working in small groups on mathematics. For example, Lampert told the students that the reason they were working in small groups was to try to understand what it was they were working on. Through all her statements about how to work together on tasks in small groups, like the small group rules, Lampert scaffolded (Vygotsky, 1978) students throughout the year in learning what it is like for mathematicians to communicate with each other about mathematics. The first day's task described above – comparing answers – was, relatively speaking, cognitively easy. Students were asked to begin communicating with each other through sharing. Later in the year, students were expected not only to share ideas with each other but to argue and convince each other and come to agreements – tasks that were cognitively more complex.

One of the most important aspects of Lampert's work with small groups in terms of learning opportunities was her emphasis on interaction among the students in the groups. If students are not allowed to interact within the group but are instead just seated as a group, then the prescriptions of the cognitive change literature are not applicable, because interaction is critical in producing cognitive change. Theorists in the area of cognitive change who acknowledge social components in the learning process all theorize that social interaction is crucial if a child, or any person, is to learn (e.g., Vygotsky, 1978; Wertsch, 1986; Bell, Grossen, & Perret-Clermont, 1985).

By early November, Lampert became more demanding in her expectations of student work in small groups. She told the students to work independently on the problem of the day and then talk with others at the table to come up with an answer on which all group members agreed. This task was quite a change within a short period of time from students' just sharing ideas with each other. Having to come to a group decision requires that students give clear explanations of their ideas so that others can then decide whether or not they agree. Lampert helped several students during group time to accomplish this new goal. She told one student to explain his answers in a way that everyone at the table, including the teacher herself, would understand and then see if everyone agreed with him. In another instance, Lampert asked a student why he had decided to use division in a problem he was trying to solve. The student told her he was using division because the group had told him to. Lampert replied that just doing what someone tells you to do, especially if it doesn't make sense, is not a good way to do math – you do not learn anything doing it that way. She said that what counts in mathematics is whether something makes sense to you. Her statement made it absolutely clear that getting the correct answer is not the most important thing in mathematics and that it is a misuse of group resources to simply copy someone's solution to a problem without understanding the reasoning behind it. Several theories of how students learn include giving explanations as one of the key components of learning from peers: Noreen Webb's work on small group learning (e.g. Webb, 1983), Wittrock's (1974) model of generative learning, and Bargh and Schul's (1980) work on cognitive benefits of teaching.

Shortly after requiring students to come to some kind of agreement in their groups, Lampert went even further in her expectations of students during small group time by telling the class that if the group agreed on an answer, each student in the group would have to be able to explain it. From these comments one recognizes Lampert's adjusted expectations of students' work as it became clear to her that they were able to handle more responsibility.

A few days later, the teacher again reminded the class that the groups were supposed to work together to come up with an explanation and that students who did not understand the problem were expected to ask questions until they did understand. Lampert said that it was the individual responsibility of each student to make certain that he or she understood. Some students might be tempted to opt out of the group work and then lay the responsibility for their lack of understanding on the other group members by saying that the others had not explained the problem well enough. By making the students individually responsible for their own understanding, however, the teacher made it impossible for students to blame their lack of understanding on anyone else.

The teacher continued through the year to scaffold students about how to work together. From February through the end of the year, the teacher told students to work together in a variety of ways; again, she was always concerned that the students talked to each other in ways that made sense and listened to each other. For example, in one instance when a student showed Lampert his idea for solving a problem and asked if he could do it his way, she told him that he could do it that way as long as it made sense to him and he could explain it to the other people in his group. Another day she instructed the students to work with partners to figure out if the pattern they had been talking about worked or not – whether it made sense – to come to an agreement about it, and to figure out what they wanted to say about it in their notebooks. On this day she added that students were to do their experiments with their partners.

In the final curriculum units, the expectations of small group work were far higher than those at the beginning of the year that simply asked students to share their ideas with one another. By the end of the year, students were communicating with each other in sophisticated ways, making it possible for students to encounter and develop ideas with their peers, a process they had not engaged in at the beginning of the year.

Students were not generally required to talk with each other as they began work on the problem of the day. There were instances, however, when Lampert told the students they were to work together on a particular problem. In addition to telling students to work together, the teacher often structured the lesson in such a way that students were required to work together. For example, one day in March the problem of the day asked students to agree or disagree with the conjecture written on the board, to write their reasoning in their notebooks, and to try to convince the group about their reasoning. At other times, the students were required to share the class materials, and thus they had to work together on a problem. This was especially true during the fractions unit when students worked with fraction bars. Each pair of students was given one set of fraction bars to work with.

As a collection, Lampert's comments to the class about how to work together on academic tasks reflect a concern for and a valuing of grappling with content, trying to make sense of mathematics both by explaining one's own ideas and by listening carefully to others' ideas and attempting to understand them. In addition to talking about how students were to work cognitively with each other while in small groups, the teacher also talked to the class about how they should work together socially.

How to Work with a Partner: Social. In her description of the small group rules at the beginning of the year, Lampert told the students that they were responsible for their own behavior and were expected to help anyone in the group who asked for help. It is interesting that these social rules about how to behave in a group were actually the first two rules about working in groups. This may indicate that the teacher believed that students have useful and interesting things to talk about with their peers, but unless there is some organization or social expectation, students will not benefit as much from their discussions as they might otherwise.

Other comments in this category do not show up in the transcripts until early November. Perhaps this indicates that up to this point Lampert had been concerned with cognitive skills, that is, getting the students to talk about mathematics with each other and giving reasons and explanations. As students became more familiar with the idea of communicating about math, the teacher added the expectation that students learn the social skills for interacting with other people. This seemed to be the case because the first thing the teacher said to the class about communicating socially in groups was that she had noticed that some people got carried away and frustrated while working in their groups and that some people needed to work on their social skills.

It appears that there was a rhythm in the class that alternated, like a pendulum, between an emphasis on cognitive skills and an emphasis on social skills. The swing was perhaps triggered by the teacher's perceptions of students' difficulties in either realm. As mentioned above, the teacher chose to spend time talking with the students about the social skills required in small groups after it became apparent to her that students were having social difficulties in one of the small groups.

Later the same day that Lampert had mentioned that some groups were getting frustrated during class, she told one student that doing what someone told him to do when it did not make sense to him was not a good way to do math. She suggested that it might help the whole table make sense of the problem if they made a diagram. The teacher was telling the student that in doing mathematics, the social rule of doing what someone tells you to do does not apply, that it is more important that something make sense to you. Again, this seems to indicate that the teacher, while concerned with the students' social skills in working in small groups, was also concerned that students were doing mathematics that was reasonable to them.

Just a few days later the teacher reported to the class that she had told a student to explain his answer to only one other student in the group because the group was just beginning to figure out how to work together. Lampert seemed to be implying that learning to work together required not only having explanations and reasons for ideas but being able to communicate those ideas and reasons to others in the group in such a way that the group members both listen to and understand what is being shared with them.

In early February Lampert told the class that she had noticed some pairs of students had been distracted from work. She said she did not want to tell these pairs not to work together because then they would not learn how to work with each other. Instead, she had decided to let them figure out how to work together, but she also wanted them to get the work done. In this instance the teacher clearly indicated her concern that students learn to work with different kinds of students in their small groups. She was willing to give the students another chance at learning to work with their current partners, even though some were having difficulty in getting mathematics work done. She was not neglecting the cognitive aspects of group work, but she seemed to be just as concerned that students develop their social skills in group work. At this point, the pendulum crested on the social side. The trade-off is evident. From a cognitive perspective, giving the students more time in a "dysfunctional" group is simply wasting time; no quality learning is taking place. From a social perspective, however, the additional time spent on group dynamics is crucial and necessary.

From Lampert's point of view, both types of learning are important and to neglect one for the sake of the other does a disservice to the students in their future work in mathematics and with other people. Research into the social dynamics of communication has revealed that students working together in small groups behave differently depending on who is in their group, specifically in terms of social status and gender. For example, Elizabeth Cohen (1972) has found that students of color are significantly less likely to talk when there are white students in the same group, and this is particularly salient when the group includes white boys. Fortunately, she has also found that such inequities can be diminished as students spend more time working in small groups with a variety of different kinds of students. Through Lampert's decision to have her students work with a variety of other students during the course of the year, she encouraged an increased tolerance and diversified acceptance among students as they worked in small groups.

Another body of research also supports the practice of heterogeneity in group composition, specifically in regard to ability. Social constructivists, basing their work on the unique approach of Vygotsky (1978), have found that students often make significant leaps in understanding when working with peers who are within their zone of proximal development

(ZPD), where ZPD represents the range of cognitive change in which a student is ready to learn with a relatively small amount of instruction. In other words, a student who has the opportunity to work with someone who has a somewhat better understanding of the concept under discussion is more likely to learn than a student who is working with someone of similar ability and understanding about the concept. By providing opportunities for her students to work with a variety of other students, Lampert increased the probability of heterogeneous ability grouping and thus the possibility that a student would be working with someone who was within their zone of proximal development.

Later on in the year, the teacher noticed one student who was working with a different small group than the one to which she had been assigned. The teacher told her to go back to her own group. This stress on remaining in your own small group again reflected the teacher's belief that it is important to learn to work with different kinds of people and that it is not okay for students to desert their groups and go off in search of others to work with if they are having difficulties in working with their own group.

As mentioned above, on one day the students were asked to find as many tangrams as possible that met certain conditions in their graphs. The students became highly engaged with this problem and began comparing the number of tangrams each group found. The teacher interrupted this activity to tell the class that the goal of the day's activities was to learn about shapes, not to be competitive between groups. Again her concern with learning mathematics was connected to her concern that students learn how to work together.

Although the students appeared to become more competent in their social skills in working in groups over the year, they had not become experts in their communications. This was evident in a teacher's comment on the third-to-last day of school that it was not appropriate to try to convince people by raising one's voice or banging on desks. Clearly there was still some room for improvement although the students had come a long way from the beginning of the year.

Evaluation of Group Work. In addition to scaffolding group work, Lampert provided feedback to the groups about their work together, both cognitively and socially. These evaluative comments were made consistently through the school year from early September through June, and as with the directions Lampert gave the class about working together, one can again see the rhythm between the cognitive and the social. The evaluative comments fall into two subcategories: comments made to specific groups or individuals within groups, and comments made to the class as a whole. Most of the teacher's comments were positive – she praised the groups for working well together – although her comments were occasionally intended to be corrective, pointing out where particular groups needed additional work.

Early in the year, soon after the class began working in small groups, Lampert stopped at one of the small groups and complimented them for doing an excellent job "collaborating" that day. She then told them that collaborating means working together and helping other people develop their ideas. She did not praise the students because they had gotten the right answer or because they had worked quickly on the task. Instead, she told the group that the thing she liked most about the way they worked together was that they were helping each other develop their ideas.

The small groups did not always work well together, and the teacher pointed out to particular groups that she was aware of their behavior and did not approve. For example, Lampert told one group that they were disrupting the class and behaving inappropriately. She said that none of the three group members had made a contribution to the class discussion and that she did not have any evidence they were trying to learn something. It is interesting that although the class was involved in a whole group discussion at the time, the teacher still addressed the students by referring to the small group and not just the individual students. By doing this, she seemed to be telling the class that she saw each individual as part of a small group unit and that the behavior, both good and bad, of the group members reflected on one another.

Other groups reaped Lampert's displeasure on occasion. At the end of November, she told a couple of boys that they had not done a good job being actively involved in the lesson. A couple of days later, another group was told that they did not look productively engaged in the work. In both cases the teacher emphasized that what she did not like about their behavior was that they were not as involved in the lesson as they should be. She did not reprimand them for talking or disturbing others or just goofing off. She specifically told them that she expected them to be fully engaged in the class and the lesson content.

Sometimes the comments that Lampert made to the class were more general in terms of her evaluation of their work. These general comments were almost always positive – usually stressing how well the students worked together during the lesson. For example, one day in November, Lampert told the class they had done a good job working that day –

paying attention to the lesson and working with each other. On several occasions she specifically mentioned collaboration in her praise. In the middle of May Lampert told the class that there had been many varieties of excellent work going on in class that day with really good collaboration, serious attempts to figure things out, and serious attempts on the part of many students to explain their thinking to each other, to the teacher, and even to other groups.

It is noteworthy that Lampert did not tell her students she was pleased with them because they had gotten the right answer or because they had been quick to solve the problem. Instead, she told them she was impressed with their attempts at communication, both in explaining ideas and trying to figure out what someone else was saying. It is important to notice that Lampert praised students for working together in the ways that she asked them to work throughout the year. In other words, she was consistent in her expectations of student work in small groups. Unfortunately, it is not always the case that what a teacher says he wants and what he actually praises students about are the same thing. In Lampert's class, the students were clearly told all year long that they were expected to work hard at making explanations and trying to figure out what others were saying – in general, to make sense of the mathematics they were doing. Consistent with her demands, Lampert called attention to and praised behavior that met with her expectations. Lampert's formal evaluation of the students, grading, also closely followed her expectations of the students. At one point during the year when she was handing back graded quizzes, the teacher specifically told the class that part of their mathematics grades was determined by what they did in their small groups and was not solely determined by their quiz grades.

Lampert made many nonverbal moves to show that she valued the work that went on in the small groups. Like most teachers, Lampert was acutely aware of the passage of time in her class (Ben-Peretz, 1990), but she gave the students time to work in their groups almost every day and she spent quite a bit of class time talking about small groups. On occasion, she even chose to use the entire class time for students to work in their small groups instead of bringing them together for a whole class discussion. On two of these occasions she specifically told the class that she had decided not to have large group discussion because most people were very productively engaged in work with their partners. Again, the rhythm between the cognitive and the social becomes evident in that although Lampert was clearly concerned with the students working well together in the small groups, she was also concerned that the students had opportunities to learn to work with different people.

Conclusion

This classroom is an example of what a teacher can do to create and maintain effective collaborative small groups through her talk to her class. This case study suggests that getting students to engage in sophisticated, complex discussions about mathematics is a yearlong endeavor, from assigning students to small groups, to creating meaningful tasks for students to talk about within their groups, to maintaining a climate in which a spirit of meaning seeking is clearly more important than finding the "right" answers.

The teacher's role in small group work is an area that deserves more investigation. Studies of other teachers in other classrooms would provide us with a better sense of how the teacher influences the interactions and learning that occur within the small groups and why some forms of small group teaching are more effective than others. How much of what goes on in the groups is a function of what the teacher says? What other teacher factors influence learning within groups? One aspect of the teacher's role in small groups that deserves further study is the challenge of participating in the small group discussions. Because she cannot be privy to all conversations within the groups, the teacher is often put in the position of dropping in and out of conversations. Studies have found that this aspect of group work can provide both benefits and difficulties for students (e.g., Blunk, 1996). In addition to direct teaching about how to work in small groups, a teacher is able to provide indirect instruction through her own modeling of the kinds of interactions that she expects to see within the small groups. For example, a teacher can set the tone for a caring, concerned approach to addressing others and their ideas by engaging in modeling, dialogue, and the practice of a caring community (Noddings, 1988). It would be useful to investigate the relationship between such direct and indirect instruction about small groups. The role of the teacher in creating and maintaining effective small group interactions is a rich area for further research.

Acknowledgments

The research reported in this chapter was supported by the Spencer Foundation under grant no. 1992–00088 with additional support from the National Science Foundation under grant no. TPE–8954724. The opinions expressed do not necessarily reflect the views of either foundation.

Note

1. This transcript was selected to use here because of the good audio quality of the discussion on the videotape, but it is certainly not the only conversation of such quality with a small group.

References

Bargh, J. A., & Schul, Y. (1980). On the cognitive benefits of teaching. *Journal of Educational Psychology, 72,* 593–604.

Bell, N., Grossen, M., & Perret-Clermont, A. (1985). Sociocognitive conflict and intellectual growth. In M. Berkowitz (Ed.), *Peer conflict and psychological growth* (New Directions for Child Development, no. 29), pp. 41–54. San Francisco, CA: Jossey-Bass.

Ben-Peretz, M. (1990). *The teacher–curriculum encounter.* New York: SUNY Press.

Blunk, M. (1996). *The role of cognitive mechanisms and social processes in cooperative learning.* Unpublished doctoral dissertation, University of Michigan, Ann Arbor.

Brown, A., & Palincsar, A. (1982). Inducing strategic learning from texts by means of informed, self-control training. *Topics in Learning & Learning Disabilities, 21,* 1–17.

Cobb, P. (1995). Mathematical learning and small group interaction: Four case studies. In P. Cobb & H. Bauersfeld (Eds.), *The emergence of mathematical meaning: Interaction in classroom cultures.* Hillsdale, NJ: Erlbaum.

Cohen, E. (1972). Interracial interaction disability. *Human Relations, 25* (1), 9–24.

Gerleman, S. (1987). An observational study of small-group instruction in fourth-grade mathematics classrooms. *Elementary School Journal, 88* (1), 3–28.

Good, T., Grouws, D., & Mason, D. (1990). Teachers' beliefs about small-group instruction in elementary school mathematics. *Journal for Research in Mathematics Education, 21* (1), 2–15.

Good, T., Grouws, D., Mason, D., Slavings, R., & Cramer, K. (1990). An observational study of small-group mathematics instruction in elementary schools. *American Educational Research Journal, 27* (4), 755–782.

Johnson, D., Johnson, R., & Skon, L. (1979). Student achievement on different types of tasks under cooperative, competitive, and individualistic conditions. *Contemporary Educational Psychology, 4,* 99–106.

Kantor, R., Elgas, P., & Fernie, D. (1993). Cultural knowledge and social competence within a preschool peer culture group. *Early Childhood Research Quarterly, 8,* 125–147.

Lampert, M., & Ball, D. (1990). *Mathematics and teaching through hypermedia* [Computer database lesson transcript, 1990]. Ann Arbor, MI: M.A.T.H. Project [Producer].

Lave, J., & Wenger, E. (1991). *Situated learning: Legitimate peripheral participation.* New York: Cambridge University Press.

Lindquist, M. (1989). Mathematics content and small-group instruction in grades four through six. *Elementary School Journal, 89* (5), 625–632.

Noddings, N. (1988). An ethic of caring and its implications for instructional arrangements. *American Journal of Education, 96,* 215–230.

Palincsar, A. (1986). The role of dialogue in scaffolded instruction. *Educational Psychologist, 21,* 73–98.

Palincsar, A., & Brown, A. (1984). Reciprocal teaching of comprehension-fostering and comprehension-monitoring activities. *Cognition and Instruction, 1* (2), 117–175.

Palincsar, A., & Brown, A. (1986). Interactive teaching to promote independent learning from text. *The Reading Teacher, 39* (8), 771–777.

Russell, J., Mills, I., & Reiff-Musgrove, P. (1990). The role of symmetrical and asymmetrical social conflict in cognitive change. *Journal of Experimental Child Psychology, 49,* 58–78.

Swing, S., & Peterson, P. (1982). The relationship of student ability and small-group interaction. *American Educational Research Journal, 19,* 259–274.

Vygotsky, L. (1978). *Mind in society.* Cambridge, MA: Harvard University Press.

Webb, N. (1983). Predicting learning from student interaction: Defining the interaction variables. *Educational Psychologist, 18* (1), 33–41.

Wertsch, J. (1986). *The social formation of the mind: A Vygotskian approach.* Cambridge, MA: Harvard University Press.

Wittrock, M. C. (1974). A generative model of mathematics learning. *Journal for Research in Mathematics Education, 5,* 181–196.

Yackel, E., Cobb, P., & Wood, T. (1991). Small group interactions as a source of learning opportunities in second grade mathematics. *Journal for Research in Mathematics Education, 22,* 390–408.

9 Teaching and Learning Politeness for Mathematical Argument in School

Peri Weingrad

> If you disagree, like Anthony just disagreed with Eddie, that's very very important to do in math class. But when you disagree, or think somebody misspoke, you need to raise your hand and say, I think he must have meant plus not times. And then Eddie will probably revise, even before you get it out of your mouth... So one thing we have here is, how to challenge or disagree with somebody in your class.. And that's a very important thing, mathematicians do it all the time. But you have to have a good reason, and you have to do it with politeness.
>
> – Magdalene Lampert on September 11, 1989, to her fifth-grade mathematics class

Confronted by a stark contrast between the expert and novice views of the work of academic disciplines and projections of the diminished stature of the United States in the global economic arena, reformers of U.S. mathematics education have begun to call for classroom discourse to more closely resemble the professional discourse of mathematicians. Contemporary reformers have a dual objective in their focus on discourse: Discourse is valued both as a learning activity and as a means of "practicing" an academic discipline. More specifically, in its *Professional Standards for Teaching Mathematics,* the National Council of Teachers of Mathematics maintains that discourse in which all students reason and argue about mathematical meanings should become a classroom norm.[1]

For classroom "practitioners," teachers and students alike, accomplishing the aims of the *Standards* is fraught with personal and social risk. David K. Cohen (1988) calls this kind of teaching and learning practice *adventurous* because it "opens up uncertainty by advancing a view of knowledge as a developing human construction and of academic discourse as a process in which uncertainty and dispute play central parts" (p. 58). What generates the risk of adventurous practice is the collision of the reformers' vision with a widespread, long-standing view of the nature of mathematical knowledge and the norms of discourse that teachers and students bring to their classrooms: The traditional view of mathematical knowledge is that it is objective; the prevailing nature of discourse is nonadversarial. The traditional view and the prevailing practice are mutually reinforcing: Together they serve to devalue reasoning and arguing and the rhetoric characteristic of disciplinary expertise in, for example, mathematics.[2]

The authors of the *Standards* acknowledge that adventurous teaching and learning involves intellectual risk taking and recognize that risk is an impediment to reform, yet they are optimistic that the obstacle may be overcome in the way teachers treat students' ideas:

> Students are more likely to take risks in proposing their conjectures, strategies, and solutions in an environment in which the teacher respects students' *ideas,* whether conventional or nonstandard, whether valid or invalid. Teachers convey this kind of respect by probing students' thinking, by showing interest in understanding students' approaches and *ideas,* and by refraining from ridiculing students. Furthermore, and equally important, teachers must teach students to respect and be interested in one another's *ideas.* (pp. 57–58, italics added)

What "respect" and "interest" actually look like in classroom discourse when applied to ideas is not specified.

Of course, students are expected to respect one another and their teachers in most classrooms. Teachers are supposed to make lessons interesting and good students are interested students. But such traditional conceptions of respect and interest are more closely aligned with issues of self-esteem and motivation than with the character of intellectual work. Discourse in conservative classrooms that exemplify the traditional conceptions of respect and interest may radiate a "niceness," a nonjudgmental aura, neither positive nor negative, that makes for intellectual relativism (Morine-Dershimer, 1983). That all ideas are respected or found interesting does not imply that all ideas are correct.

To assess and evaluate the reforming of classroom discourse, a relevant analysis of either teacher or student talk must distinguish that which is simply respectful of students' ideas and intended to protect their self-esteem from that which is respectful yet concerned with the relationship of students' ideas to mathematical investigation. Such projects require explicit formulations of how adventurous teachers convey respect for students' ideas and how they teach students to respect and be interested in one another's ideas. To know whether these teachers are successful also requires explicit formulations of how students who have learned to respect and be interested in each other's ideas interact. Such formulations are not evident in the literature of the field. Hence, the project of the present work is the development of provisional criteria for distinguishing intellectually adventurous classroom discourse from the conservative. The approach involves the identification of different kinds of risk, some of which may be found in conservative classroom lessons and some of which may be peculiar to adventurous lessons.

Risk for Face

Taking risk as the dimension along which to distinguish adventurous from conservative teaching and learning is a more complicated tack to take than it might first appear because risk is not indigenous to reformed classrooms; on the contrary, risk accompanies conservative practice as well. Erving Goffman (1955), the early advocate of the view that risk inheres in all social interaction, maintains that particular forms of language and discourse are designed for the potential dangers of communication. He broadly describes the kinds of risk involved:

> The structural aspect of talk arises from the fact that when a person volunteers a statement or a message, however trivial or commonplace, he commits himself and those he addresses, and in a sense places everyone present in jeopardy. By saying something, the speaker opens himself up to the possibility that the intended recipients will affront him by not listening or will think him forward, foolish, or offensive in what he has said. . . . Furthermore, by saying something the speaker opens his . . . recipients up to the possibility that the message will be self-approving, presumptuous, demanding, insulting, and generally an affront to them or to their conception of him. (pp. 227-228)

What it is that Goffman characterizes as being in jeopardy in social encounters is, in the literature, variously and technically termed *face, identity,* or *self.* For Goffman,

> Face is an image of self delineated in terms of approved social attributes – albeit an image that others may share, as when a person makes a good showing for his profession or religion by making a good showing for himself. (p. 213)[3]

In principle, then, risks to face and obstacles to communication are found in every classroom, adventurous and conservative alike.

Distinguishing between adventurous and conservative discourse along the lines of risk presumes that the risks of each are different in kind or in number. The presumption here is that because the aim of reformers is to put new views of knowledge and of academic discourse into practice, the difference is qualitative. Regardless, that there is any discourse at all in the "face" of the obstacles in both kinds of classrooms deserves explanation.

Penelope Brown and Stephen Levinson's (1987) theory of *politeness* is an effort to explain how it is that social interaction is ubiquitous in spite of the face-threatening acts (FTAs) inherent in communication. Brown and Levinson posit that speakers' overt endeavors to maintain the face of their FTA-receiving interpreter is what ensures the recipients' ongoing cooperation in face-threatening activities.[4] Their theory accounts for their identification of particular linguistic forms as embodying *politeness strategies,* which are the means by which speakers redress threats to their interpreters' face. Politeness strategies are designed to counteract the potential damage of FTAs by indicating that their speakers generally do not intend to threaten their interpreters' face. Accordingly, a teacher whose students reason and argue about (mathematical) meanings is a teacher who redresses the threats to face inherent in reformed, adventurous classroom practices. Such a teacher may be described as using politeness strategies. The premise of this investigation is that the constructs of politeness theory may usefully be appropriated by researchers of classroom discourse.

Brown and Levinson identify two broad varieties of politeness strategies, *positive* and *negative,* and infer that each kind is designed to maintain a corresponding aspect of an interpreter's face. The contrast of positive and negative is not parallel to that of normative descriptors of discourse strategies such as "good" and "bad" or "productive" and "unproductive." Each term refers to the aspect of face a strategy addresses: Negative politeness is FTA redress that aims at maintaining an interpreter's negative face whereas positive politeness aims at maintaining an interpreter's positive face. Negative politeness is evidence that the speaker does not intend to restrict his interpreter's freedom of action; positive politeness is evidence that the interpreter's beliefs, desires, or intentions are shared, and likewise valued, by the speaker (for brevity, future reference will be made solely to the sharedness of beliefs). Consequently, FTAs that restrict the interpreter's freedom are threats to negative face (negative FTAs); FTAs that deny that the speaker shares the interpreter's beliefs are threats to positive face (positive FTAs). For Brown and Levinson, positive and negative face threats are not mutually exclusive categories: Threats to positive face may also be threats to negative face, because for a speaker to indicate that he does not share some of the beliefs of his interpreter may be to deny that they share the belief that the interpreter should have his freedom (p. 67).

No significant strain results from drawing an analogy between each of the negative and positive varieties of politeness and, respectively, the respect and interest the authors of the *Standards* maintain are required for all students to reason and argue about mathematical meanings. Brown and Levinson themselves identify negative politeness with a folk notion of respect and expressions of restraint and positive politeness with expressions of *solidarity* (pp. 132–142). Solidarity may be defined much like *interest* – as having a stake in another's affairs.

Politeness theory is used here to frame a treatment of Magdalene Lampert's teaching practice in a lesson where the discourse is congruent with the practices advocated in the *Standards.*[5] The purpose is to propose and investigate a methodology for research that ultimately could justify claims about teachers' teaching and students' learning to respect and be interested in other students' ideas. Different kinds of FTAs and the politeness strategies associated with Lampert's questioning, criticizing, and challenging her students are examined in developing provisional criteria for distinguishing risk and redress peculiar to adventurous lessons from that which may also be found in conservative classrooms.

Face-Threatening Acts in the Lesson

One reason for studying the risk and redress in a lesson, as opposed to, for example, interactions among small groups of students, is that the lesson seems the most common and best documented type of classroom participation structure at all grade levels.[6] Another reason is that, given the assumption of the reformers that students typically enter (and leave) classrooms unprepared to respect and be interested in one another's ideas, if students do leave a classroom reasoning and arguing, the conclusion may be drawn that they have learned to do so as a result of the teacher's teaching. Therefore lessons where the teacher is integrally involved in the pattern of interactions are likely to be a fertile site for studying the reforming of classroom discourse.

The communicative rights and responsibilities of teachers and students in lessons are asymmetrical.[7] The unequal communicative rights are manifest in, for example, the teacher's determining whether to address a single student or a group, whether students must bid for the *floor* (the right to speak), whether appropriate responses are to be made individually or in chorus, whether responding is considered voluntary or compulsory, what the topic is, and which student responses are correct. Students characteristically have the right and the responsibility to speak only when sanctioned by the teacher. Although student participation in the lesson economy most typically involves speaking, sometimes merely demonstrating a desire to speak is sufficient, as when many students bid for the floor in order to answer a teacher's question.

Edwards and Furlong (1978) articulate a commonsense explanation for lessons' participation structure:

> In classrooms, it is their position as knowledge experts which justifies teachers in "owning the interaction." To the extent that their expertise is acknowledged, they will be expected to do most of the talking themselves, and to evaluate what is said by others. (p. 24)

A consequence of the teacher's "owning the interaction" is that all members of such a classroom community are aware that their action (or inaction) is interpreted with reference to the norm determined by the teacher. Goffman (1981) describes the secondary status of a common kind of student talk, an answer to a teacher question, as more dependent on the question than the question is on the answer: "Whatever answers do," he observes, "they must do this with something already begun" (p. 5). Thus student action is often best interpreted as responsive to teacher initiative.

The participation structure of the lesson systematically engenders certain varieties of both negative and positive FTAs. Two classes of each kind are investigated here: *Request for Bids* and *Nominations* are negative FTAs that occur when students interpret the teacher as commanding or requesting that they display their desire to speak and that they actually speak, respectively; *Criticism* and *Challenge* are positive FTAs that occur when students interpret the teacher as not sharing their beliefs. On Brown and Levinson's account, to maintain students' negative face so that they will continue to participate involves the teacher's expressing restraint in her requests that students volunteer or her commands that they speak; to maintain students' positive face involves the teacher's indicating that she shares some of the beliefs of the students.

Regardless of how a student responds, Requests for Bids and Nominations may threaten students' negative face because of the power asymmetry in classrooms. The interpretation of acts such as these as commands or requests involves the interpreter in imputing the expectation to the speaker that the interpreter will proceed to act relevantly with reference to the command or request. Commands and requests threaten interpreters' negative face because they indicate that the speaker intends that the interpreter's freedom of action be constrained (to responding as expected). What counts as relevant in a lesson is, as argued above, typically determined by the teacher. Following a teacher Nomination, for example, everyone will interpret students' subsequent action (or inaction) as responsive to the teacher's initiative act: A student may either respond substantially as expected or not. Responding as expected opens the student up to the possibility that others will affront him by not listening or will think him, for instance, "wrong" therefore "dumb in math" or "right" therefore "smart in math" (which is not always, in a typical American classroom, a compliment). Responding unexpectedly, whether irrelevantly or by opting out, opens the student up to the possibility that the others will think him "dumb in math" or uncooperative. To the extent that a student is concerned with what the others think of him, his knowledge of the possible consequences constrains his freedom. In a lesson where the discourse is congruent with the vision in the *Standards*, engaging in reasoning and arguing about mathematical meanings may be a risk for students' face sufficient to deter some from participating (Jackson, 1968; Lampert, Rittenhouse, & Crumbaugh, 1996).

The implications of the teacher's owning the interaction of the lesson for the positive aspect of face may similarly be deduced: To the extent that students value the teacher's esteem, an act that reveals the teacher as critical of or in disagreement with students' beliefs is threatening to students' positive face. The Criticism encountered here relates to the students' beliefs about their own communicative rights and responsibilities in a lesson, and it may be a threat to positive face because Criticism is evidence that the teacher does not share those students' beliefs. Challenge involves the teacher's questioning the meaning of something a student says and may be a threat to positive face because questioning is an indication that the teacher does not find what the student says wholly acceptable and is evidence that the teacher does not share all the beliefs of the student. As noted above, positive FTAs may also be threats to negative face; but, for the sake of brevity, this discussion of Criticism and Challenge refers only to their positive face-threatening aspect.

Given the goal of distinguishing discourse that is simply respectful of students' ideas from that which is also concerned with the mathematical character of those ideas, it might be suspected that the negative FTAs of Requests for Bids and Nominations are found in both adventurous and conservative lessons, while the positive FTAs of Criticism and Challenge – since they concern beliefs and therefore ideas – are indigenous to adventurous lessons alone. But this is not the claim being advanced; rather, the suggestion here is that Criticism is a kind of positive FTA that may be found in all lessons, whereas Challenge may be a risky practice that distinguishes the adventurous from the conservative. There are varieties of positive FTAs that are common to traditional and adventurous lessons because the practitioners of each kind may be at odds over beliefs about behavioral issues (e.g., the means of class management) or over beliefs about intellectual issues (e.g., what the correct answer is). Whether the beliefs in question concern the behavioral or the intellectual, any act interpreted as the teacher's being critical of or in disagreement with a student's beliefs is threatening to the student's positive face. What is expected to distinguish adventurous lessons is therefore not the simple fact of positive FTAs, but their content. What is implied by the vision of the reformers is that the positive FTAs of adventurous mathematics lessons will revolve around reasoning and arguing about mathematical meanings and not around the evaluation of student answers, which is the concern of traditional, conservative lessons.

What follows is a first attempt at characterizing more narrowly the behavioral grounds for distinguishing negative FTAs and Criticism from the more intellectually oriented Challenge. The strengths and weaknesses of the constructs of politeness theory will be examined, specifically in regard to their adequacy for tracing the linguistic expression of respect and interest. To ground this investigation, a description of the lesson under study is presented directly below. Excerpts from a transcript of this lesson are used to explore the extent to which the concepts of face, face-threatening acts, and redress can illuminate our understanding of reforming the discourse of mathematics lessons.

The Classroom Record: September 11, 1989

The object lesson occurs at the very beginning of the school year, on September 11, 1989.[8] Lampert and the students had been introduced to one another on September 7, the first day of the 1989–90 school year, but the eleventh is the first day that Lampert and the students have what the regular classroom teacher, Thom Dye, refers to as their "math time" together.[9] Because this day is at the very beginning of the instructional year, the normal daily routine has not yet become established, and the lesson is interspersed with short periods of individual student work. The motive behind the choice of this lesson is that it provides what may be the best opportunity to observe the students and teacher acting according to expectations and intentions uninformed by previous interaction. The circumstances may lend insight into the students' preparedness to be respectful of and interested in each other's ideas.

Following lunchtime recess on September 11, the fifth graders reenter the classroom with some commotion, and Dye rings a small bell to draw their attention. The students take a minute or so to get into their seats and to quiet down, then Dye reintroduces Lampert. After a few more words about classroom procedures, Dye cedes control to Lampert, who continues the practice of beginning with introductions as she names the observers and videographers for the students. Greeting routines out of the way, Lampert proceeds with the lesson.

The lesson may be segmented into nine Topically Related Sets of exchanges, as follow:[10]

(1) *Revision* 06:18–14:43
Lampert first asks the students to personalize their mathematics notebooks using the black magic markers that the students also have on their desks. The students respond appropriately, taking up the markers to write their names on their notebooks. Lampert then says a few things about how she intends notebooks to function over the course of the school year. In the course of giving this information, Lampert introduces "the process of revision" as "in mathematics, one of the things that is really important to learn about" and proceeds to elicit from the students their ideas "about what revision might mean."

(2) *Black Magic Markers* 14:46–16:34
Lampert explains that what she has been telling the students so far are the reasons for the presence of the tools – the black magic markers and the notebooks – and works out the logistics for the assignment of one of the students as a "math aide" who will "collect the magic markers every day and then pass them back out again for math."

(3) *Notebooks* 16:35–17:33
Returning to the other tool, the notebooks, Lampert asks the students to "tell [her] what they notice about it," and the class discusses what the graph paper, which makes up the pages of the notebooks, is and what its function might be.

(4) *Horizontal and Vertical* 17:34–28:48
Describing the lines across the paper at right angles to each other which form graph paper is the occasion for discussing "horizontal" and "vertical," "two important new words you're going to learn this year."

(5) *Doing Math* 29:04–37:31
Having finished with the discussion of graph paper, Lampert describes what will be the usual course of their math lessons and

instructs the students to begin to practice what will become routine. The students comply by writing in their notebooks as directed and begin to work on the problem of the day.

(6) *Between* 37:31–40:57
Once Lampert has satisfied herself that the students have copied the problem of the day into their notebooks and done some work on it, she writes on a chalkboard "4 + 3" and "7 + 9" and says that she wants the students to tell her "another addition problem whose answer is in between the answer to this one and the answer to this one. .....Without saying the answers." Lampert proceeds to elicit such addition problems from a number of the students and writes them on the board in between the two original problems.

(7) *Why All These Work* 40:58–47:58
After collecting a number of addition problems from the students, Lampert asks the students to "tell me why all these work" and to "explain, why all of these problems have answers in between these two." Lampert obtains explanations from several students, many of which explain why a single problem "works" as a solution.

(8) *True or False* 47:59–51:05
Lampert now "take[s] account" of the students' explanations of why some solutions work by formulating a general rule, and she asks the students to decide whether her rule is true and to formulate an explanation of why they think so.

Face-Threatening Acts in the Classroom Record

To support the development of criteria for identifying the negative FTAs of Requests for Bids and Nominations, an excerpt from *Revision,* the very first set of exchanges between Lampert and her students, is transcribed below. Line numbers of an utterance preceded by an asterisk contain instances of the FTAs that are treated below in some detail.

Negative Face-Threatening Acts: Request for Bids and Nomination

Revision

(a) – 09:00

*1 Lampert: [Beginning of turn excised.] Does anybody think they have an idea to tell the class, about what revision might mean?

2 Students: [Raise hands.]
3 Lampert: Okay
4 Anthony has an idea, and Shahroukh has an idea.
5 Charlotte has an idea, and Varouna.
6 Okay, I'd like to hear all those ideas, //but,// it's not only me that you're talking to, it's everybody in the class.
7 Students: //[Lower hands.]//

(b) – 09:28
1 Lampert: And if you if you haven't raised your hand with an idea, I want you to listen to the people who have an idea what revision might be, and see if you agree or disagree, and if you have something to add.
2 Okay?
*3 Charlotte, what's your idea about revision?
4 Charlotte: Um, that you write down one idea that you think, and then you decide that maybe that's not the right, () you don't think that that's the right idea, and that you want to change it.
5 Lampert: And you use revision by writing your other answer?
6 Charlotte: Instead of erasing your first idea and then using it again, you just, it's a second answer.
7 Lampert: Okay!
8 That that's one um good way of explaining re- revision.

(c) – 10:21
*1 Let's see,
2 Students: [Raise hands.]

(d) – 10:23
*1 Lampert: Shahroukh was another person who had an idea?
2 Students: [Lower hands.]
3 Shahroukh: I think she's got my idea.

(e) – 10:28
*4 Lampert: Is there anything you would like to add to what Charlotte said?
5 Shahroukh: Well, I think revision is, um well, it's almost the same.
6 What I'm gonna say, because what I think is revision is when you have an idea: and it's not correct and you don't want to, like, mess it all up and you don't want to try to erase or cross it out.
7 But, if you write in another line, it'll be revision, rather than making it all messy.
8 Lampert: Okay, so part of what we're worried about here is, what to do when you change your mind.
9 And one thing that people get used to doing when they change their minds is erasing.
10 And what I'm going to ask you to do in math, this year, is not to erase.
11 And that's one reason why we're going to use black magic markers instead of pencils in math.

(f) – 11:28
*1 Let's see,
2 Students: [Raise hands.]

This sequence of exchanges from *Revision* illustrates several instances of each of the two kinds of negative FTAs: there are three Requests for Bids, (a1), (c1), and (f1), and three Nominations, (b3), (d1), and (e4). Provisional criteria to delimit these FTAs in this excerpt and elsewhere in the transcript of September 11 are offered below. These criteria are abstractions made on the basis of observations of the interactions during this single lesson in Lampert's classroom, though the moves they specify are quite similar to moves that have been identified and discussed elsewhere in the literature on classroom discourse (e.g., Mehan, 1979a, 1979b; Lemke, 1990). The criteria presented here are merely an early step in developing a categorizing scheme that eventually may support more precise comparisons across classrooms and lessons.

A Request for Bids is identifiable in the transcript by virtue of being an act, *RB,* that satisfies the following three criteria:

(1) *RB* is a teacher *act.*
An act, whether a spoken or nonverbal "utterance," is distinguished from a turn-at-talk in that turns are constituted by one or more acts.

(2) *RB* is addressed to a group of students.
Evidence that the act is addressed to a group of students comprises the teacher either or both:
  i. gazing, pointing, or nodding, an action that ranges over a number of students; or
  ii. making verbal reference to the group as the intended object of the teacher's request by her use of indefinite, plural, or mass terms of address or referral, including "you," "anybody," "we," or "people," and making no specific or definite reference to any particular student as the object of the Request for Bids, such as by pointing at a student or using a student's name.

(3) *RB* is succeeded directly by either:
  i. student calls for attention (e.g., utterances such as "me! me!" or "I know!" or "ooh!" or actions such as hand raising) and followed by a teacher's Nomination (see identification criteria of Nominations below); or
  ii. silence and relevant inaction on the part of the students and then *repair* by the teacher accomplished by, for example, a

repetition of the Request for Bids or an abandonment of the normal process and a Nomination of a student who has not bid.

A Nomination is identifiable in this transcript by virtue of being an act, *NM*, that satisfies the following four criteria:

(1) *NM* is a teacher act.
(2) *NM* is preceded directly by that student's bid for the floor.
A bid for the floor is typically accomplished by students' raising their hands, or making eye contact with the teacher, or vocalizing (e.g., "me! me!" "ooh! ooh!").
(3) *NM* is addressed to one particular student.
Addressing an act toward a student is accomplished by the teacher doing one or a combination of the following:
i. referring to a student by name; or,
ii. referring to the student by a singular term of address (such as the singular personal pronoun "you"); or,
iii. gazing, pointing, or nodding in the direction of a student.
(4) *NM* is succeeded directly by the student(s) addressed either:
i. undertaking relevant action and the un-Nominated students suspending (by being silent) or withdrawing (by lowering their hands) their bids to speak; or
ii. opting out of substantively responding.
Evidencc that the student is opting out comprises head-shaking, shrugging, saying "no," or being silent or otherwise inactive.

Positive Face-Threatening Acts: Criticism and Challenge

To support the development of criteria for identifying the positive FTAs of Criticism and Challenge, an excerpt from *Why All These Work,* which exemplifies the positive FTA of Criticism, is transcribed directly below. This excerpt, quoted in part at the outset, contains the sole example of Criticism in this lesson.

Why All These Work

(g) – 43:29

1 Lampert: [Beginning of turn excised.] Eddie?
2 Eddie: Um..I think if there's any number below s– um nine times seven it would be lower than that //nine times seven.//
3 Anthony: //Nine times seven?//
4 Eddie: //(Plus seven)//
5 Lampert: //Okay.//

6 Now here we have a very interesting, um, exchange, which is something I think is going to ha- happen and hope happens a lot in this class.
*7 But we need to have some ways of having this exchange..
8 Eddie said, I think anything lower than nine times seven is gonna have an answer that works.
9 And Anthony said, without raising his hand, times seven?
10 Okay.
11 Okay and then Eddie revised, and he said plus seven.
12 Okay...
13 If you disagree, like Anthony just disagreed with Eddie, that's very very important to do in math class.
*14 But when you disagree, or think somebody misspoke you need to raise your hand and say, [raises pitch.] I think he must have meant plus not times.
15 [Lowers pitch to normal level.] And then Eddie will probably revise, even before you get it out of your mouth...[coughs.]
16 So one thing we have here is, how to challenge or disagree with somebody in your class..
17 And that's a very important thing – mathematicians do it all the time.
*18 But you have to have a good reason, and you have to do it with politeness..
19 But the other thing I want to come around to is really what Eddie meant.
20 Eddie, why don't you say what – say again what you're trying to say here.
21 Eddie: Okay.
22 If there's – wait – any number below seven plus nine like an addition problem that has () seven plus nine and it would will fit, but it has, like, the numbers have to be, the numbers, well they could be lower than um four plus three.
23 Um.
24 Students: [Raise hands.]
25 Eddie: And sometimes they have to be higher.
26 Lampert: Okay.
27 So what – what Eddie is trying to do here is make a generalization, about what will always work, and he put a lot of conditions on his statement.
28 Let's hear from some other people.

In contrast to the dearth of Criticism, the lesson of September 11 includes several instances of Challenge. The following exchanges from *Why All These Work* form a representative example:

Why All These Work

(h) – 47:19
1 Lampert: Okay
*2 Ivan, what do you think about that?

3 Ivan: Like if you plus four plus three that'll make seven, and seven plus nine makes sixteen and I think you need any, like, answer that is smaller than sixteen and smaller than seven.
4 ()
5 Lampert: Okay
*6 So you're making a statement that the answer has to be smaller than sixteen and,
7 Ivan: Not smaller than sixteen but not bigger than sixteen and not lesser than seven.
8 Lampert: Okay.....

Provisional criteria to delimit the positive FTAs of Criticism and Challenge in these excerpts and elsewhere in the transcript of September 11 are offered below. As above, in the case of negative FTAs, these criteria are abstractions made on the basis of the interactions of this one lesson in Lampert's classroom. Likewise, the specified moves resemble those that have been identified and discussed elsewhere in the literature on classroom discourse (e.g., O'Connor & Michaels, 1993; Pontecorvo & Girardet, 1993). Again, these criteria are intended merely to represent an early step in developing the means to enable more precise comparisons of discourses of teaching and learning.

A Criticism is identifiable in this transcript by virtue of being an act, *CR*, that satisfies the following four criteria:

(1) *CR* is a teacher act.
(2) *CR* is preceded by an un-Nominated student act.
(3) *CR* refers to the act in (2).
(4) *CR* is succeeded directly by a Request for Bids or a Nomination of a student other than the student in (2).

A Challenge is identifiable in this transcript by virtue of being an act, *CH*, that satisfies the following five criteria:

(1) *CH* is a teacher act.
(2) *CH* is preceded directly by a Nominated student's act.
(3) *CH* is a second successive Nomination of the student in (2).
(4) *CH* refers to the student act in (2).
(5) *CH* is succeeded directly by the Nominated student's responding relevantly to the teacher Nomination.

Politeness Strategies

The criteria offered here may support the identification of four different kinds of acts – Requests for Bids, Nominations, Criticisms, and

Challenges – which Lampert addresses to her students in the lesson of September 11, 1989. According to Brown and Levinson's politeness theory, these acts of the speaker teacher are threats to the face of her interpreter students.[11] An adult and experienced teacher such as Lampert may be assumed to understand (if only implicitly) that much of what she requests and commands of her students, including the four acts under examination here, may be received by them (again, if only implicitly) as threats to their face. Politeness theory predicts that Lampert's awareness of the face-threatening quality of her acts leads to her adopting politeness strategies in attempts to maintain her students' face and ensure their ongoing participation in the classroom interactions.

Teachers may redress threats to student face at several points in a sequence of acts that comprise an FTA: roughly, before, during, and after. For instance, after threatening students' negative face with a Request for Bids, a teacher may attend to the students' positive face before she has Nominated any of them to respond, as Lampert does at (b). At (b1–4), by asserting that she values modes of student participation other than speaking and demonstrating the desire to speak ("listen [...] and see if you agree or disagree, and if you have something to add"), Lampert undermines her ground for interpreting failing to bid as uncooperative. That Lampert legitimizes listening to and thinking about other people's ideas as a means of participating is an indication that she intends to be inclusive, involving even the nonspeakers in the discourse. This consideration counts as redress to students' positive face because it carries her presumption that the students intend to cooperate and thus that they share the teacher's goals for the activity.[12] Thus, Lampert attempts to maintain the face of those who want to participate in the classroom economy but who do not bid, perhaps because they don't "have an idea [...] about what revision might mean," or because they don't want to open themselves up to the possibility of being thought "wrong," or maybe even "smart."

Instances of redress that occur in the next teacher act following a student response to a Nomination deserve particular attention, given the reformers' goal of shifting the function of mathematical classroom discourse from evaluating student answers toward reasoning and arguing. When a student responds to a Nomination in answer to a teacher's question, the teacher's next move is typically construed as an evaluative move.[13] Mehan (1979a) shows that one teacher employs one of three general strategies following a student response to a Nomination: either (1) she positively evaluates the response and then may initiate another sequence of exchanges; or (2) she negatively evaluates the response and then prompts or simplifies the question; or (3) she provides no overt evaluative feedback and *recycles* the original question for another student to answer. Mehan's interpretation is that when a student response matches the teacher's preferences, she provides a positive evaluation, but when the student response does not properly stack up, she makes either a negative or no evaluative follow-up move. Thus, to put a politeness-theoretic spin on Mehan's interpretation, when a student response is "right" or otherwise suits the teacher's preferences, such a teacher as he writes about may attempt to maintain students' positive face by indicating approval. Lampert does this at (b8), saying that Charlotte's explanation is "one [...] good way of explaining [...] revision." When a response is "wrong" or otherwise undesirable, the teacher's next move threatens students' positive face as it is construed as a negative evaluation. What Lampert does to follow up such a dispreferred student response merits more detailed attention.

When a teacher follows up a dispreferred student response to a Nomination, such as an opting out of making a substantive response, for example, the event may be termed a *breakdown* in the usual discourse pattern of the classroom. A breakdown is the suspension or the abandonment of normal operating procedure that is construed as resulting from the obtrusion of unexpected or irrelevant action into the regular flow of events.[14] Often, Lampert repairs such breakdowns by adopting a politeness strategy. One instance of repair work in this transcript is Lampert's expressing confidence in the student's ability to carry out the act called for by her FTA at (j4), cited below, following the breakdown that occurs when Ivan refuses, at (j2), to take up her Nomination. According to Brown and Levinson, such optimism is aimed to redress students' positive face by expressing Lampert's belief that Ivan wants to be considered capable and intends to fulfill his wish by indicating her belief in his capacity (p. 129).

Why All These Work

(j) – 42:35

*1 Lampert: Ivan, can you explain another one?
2 Ivan: He took mine.
*3 Lampert: You want to try?
4 I bet you can explain this one............
5 Can you?
6 Ivan: [Shakes head.]
7 Lampert: You don't think so?
8 Do you want to try to explain this one? [Makes eye contact with next student speaker.]

These exchanges with Ivan, described as a breakdown in Nomination, may also seem to fulfill the criteria of Challenge, as they incorporate two successive Nominations of Ivan, at (j1) and (j3ff), and because the second Nomination may be construed as responsive to Ivan's response at (j2), as it depends on and is relevant to Ivan's response. However, such a characterization of (j) fails to accommodate a crucial difference between the positive FTA and the negative: the location of possible breakdowns in the sequence of events which define them. Although a second Nomination of the same student characterizes a successful Challenge as well as a breakdown of Nomination, breakdown of Challenge occurs directly following the *second* in the series of Nominations of the same student, whereas breakdown of simple Nomination occurs directly following the *first* Nomination. The different kinds of breakdowns may be seen in the contrast between the breakdown at (j2) of Lampert's Nomination of Ivan, at (j1) above, and the breakdowns at (k6, k9) of Lampert's Challenge of Candice, at (k5) below.

Notebooks

(k) – 16:46
1 Lampert: Candice?
2 Students: [Drop hands.]
3 Candice: It's graph paper.
4 Lampert: Graph paper?
*5 What does that mean?....
6 Candice: [Opens her mouth, closes it, shrugs.]
7 Students: [Begin to raise hands.]
8 Lampert: Suppose I couldn't see what was inside this notebook and you needed to explain to me what graph paper was?
9 Candice: [Candice shrugs again.]
10 Lampert: How could you describe it?
11 Students: [Raise hands.]
12 Lampert: Ellie?

In the case of simple Nominations, such as that of Ivan at (j1) and Donna Ruth, below at (m2ff), Lampert's second Nominations, for example, at (j3–5) and (m7), seem intended to repair the interaction and achieve the kind of response expected in the first place. However, in the cases of Challenge it is less clear that the first student response, such as at (h3) and (k3), requires "fixing."

Between

(m) – 37:42
1 Lampert: .....Without saying the answers.....
*2 [Gazes at Donna Ruth.] What do you think?
3 Can you tell me one?
4 Students: [Drop hands.]

5 Lampert: No answers, just an addition problem.
6 Donna Ruth: Seven minus four?
*7 Lampert: Addition, it has to be addition.
8 Donna Ruth: Oh
9 Students: [Have raised hands.]
*10 Lampert: What do you think might work?
11 This is four plus three and this is seven plus nine.
12 Can you tell me one that would have an answer in between those two?
13 Donna Ruth: Nine?
*14 Lampert: Nine,.. plus what?
15 Donna Ruth: Four.
16 Lampert: You think that nine plus four would have an answer that's between four plus three and seven plus nine?
17 Okay.

In contrast to the dispreferred responses of Donna Ruth at (m6) and (m13), of Shahroukh at (d3), and of Ivan at (j2), in instances of Challenge, as with Ivan at (h3) and Candice at (k3), the first student response is a preferred response and seems interpreted as a substantive contribution. Also, in contrast to the cases of Nomination breakdown, the second Nomination of a Challenge is neither a repetition nor a recycling of the teacher question, as it is at (m10), nor a negative evaluation nor a prompt, as it is at (m7) and (m14) above, nor is it a simplification of the question. Lampert's repertoire of moves for following up a student response is rather different than the repertoire described by Mehan (1979a).

The question is: What is the function of the second Nomination of Challenge? The case sketched earlier, involving the characterization of Nominations as negative FTAs in contrast to Challenges as positive FTAs, as well as the hypothesis that Challenges are distinctive of adventurous teaching and learning, hangs on the answer to this question.

In the Challenges presented here, the relevant connection between the second Nomination and the preceding student response demands a student response, just as the original question does. But the difference between the second Nomination of broken-down Nominations and the second Nomination of the Challenges seen here implies that the response to the first Nomination of Challenge does not need fixing, as a breakdown in simple Nominations does; rather, a Challenge seems more of an exploration of the student's contribution, a questing after what the student might believe that causes him to say what he does. Questioning of this kind may well be described as an expression of interest in the student's ideas. If a characterization of the questioning in Challenges as an expression of interest in what the student thinks is an adequate description, then Challenges may be taken to function as positive politeness. If, on the other hand, Challenges are pseudoquestions, intended to expose differences between the teachers' beliefs and the students' beliefs for the purpose of evaluation, then Challenges are threats to students' positive face.[15]

Limitations and Directions for Future Work

In addition to providing the ground for contrasting the negative FTA of Nominations and the positive FTA of Challenges, the examples of (j) and (k) show that not all breakdowns are repairable, even when redress is given to face. Politeness strategies thus cannot serve as complete explanations of why a student acts one way or another. Politeness theory is presented by Brown and Levinson as a theory of production: An act is performed with a particular choice of politeness strategy based on the speaker's conception of the weight of the FTA (pp. 71–84). The theory merely accounts for Lampert's choosing some particular linguistic expression in giving redress to Ivan's face at (j4), but it provides no explanation for Ivan's refusal at (j2) to "explain another one." Thus, though the theory provides tools for characterizing interactions on the basis of assuming a causal connection that leads from a speaker's conception of the weight of his impending FTA to his production of a linguistic expression, as it stands the theory does not explicitly address the relationship between a teacher's expressions of respect and interest and her students' (re-)action.

Politeness theory as applied in classrooms begins with the assumption that the teacher recognizes that FTAs may impede students' participating in reasoning and arguing; then the theory points to the teacher's linguistic expressions of politeness. The authors of the *Standards* start with the assumption that students participate in reasoning and arguing only in a polite atmosphere. Both of these assumptions deserve better justification than they have received. Better justification will require answering the questions, first, whether students interpret some aspects of the teacher's language as politeness, and, second, whether they come to reproduce it themselves.

In the light of a goal of making claims about teachers' teaching and students' learning to respect and be interested in one another's ideas, the prediction of politeness theory concerning the production of politeness is only one piece of the required framework: A complementary framework concerning the interpretation of politeness is ultimately necessary, but it cannot be developed here. Justifiably drawing the conclusion that any particular form of teacher language is indeed a partial cause of the students' continued participation in a lesson demands an analysis of the students' behavior that shows whether and how a teacher's productions are interpreted as expressions of respect and interest by her students. Such an analysis requires evidence that the same student or group of students is affected differently by what is arguably the same FTA committed with variable linguistic expression.

In developing and applying a complementary interpretive framework in combination with the productive aspect of politeness theory, any evidence that Lampert is aware of threats to face that is relevant to the overarching goal should be evidence that is also available to her students. This implies that interviews or records other than of the public, classroom events are at best of limited, indirect utility in investigating the teaching and learning of politeness strategies.

Additional limits to the explanatory power of the politeness framework that must be taken into account in the development and application of any extension to the framework are the general problems with doing discourse analysis: Any attempt to make causal claims on the basis of an analysis of records of classroom discourse must involve making inferences on the basis of patterns of acts occurring over time. These limits are evident in the implications of the fact that an instance of politeness cannot be considered the sole cause of any act: Identifying what counts as respect and interest for the students cannot be done on the basis of a small number of events. Indeed, although it is assumed here that Lampert and the reader share an understanding of her language and what counts as politeness, identifying what choices of linguistic expression count as politeness strategies for the teacher also requires a longitudinal study.

Despite the limitations of the original statement of politeness theory and the difficulties associated with doing discourse analysis in general, the ability to identify particular sequences of events as instances of FTAs, such as Requests for Bids or Nominations, and particular acts as Lampert's redressive expressions of respect and interest has real explanatory potential. Such identifications lay the groundwork for investigating the association of teachers' linguistic strategies with outcomes in terms of student discourse. It remains to be seen whether the analytic tools of politeness theory will be developed and applied in longitudinal studies that provide insights into how teachers may manage to teach and students may manage to learn how to express respect for and interest in each other's ideas.

The focus here has been on FTAs committed by the teacher and the politeness strategies that redress them. There are two other classes of student-interpreted FTAs of interest in classrooms: students' self-addressed FTAs and student–student addressed FTAs. Following Goffman, any student act is, just like any teacher act, potentially face-threatening to one who receives it. For the aims of reform, it bears keeping in mind that teachers also encounter FTAs, both in what they do themselves (e.g., taking risks by experimenting with new discourse practices) and in what their students do. Adventurous teaching is risky for teachers as well as for students. Ultimately, to investigate whether Lampert's students or any other students indeed learn to respect and be interested in one another's ideas requires broadening politeness theory's field of application by considering the "scope" of an FTA as it may differentially affect interpreters. Student-to-student and students' and teachers' self-addressed FTAs all deserve study as potentially conserving forces of the classroom status quo.

Acknowledgments

The author is grateful to the many people who have given their patient and knowledgeable support to this project. Special thanks are due Deborah Keller-Cohen, Magdalene Lampert, Mary Catherine O'Connor, and John Swales for most of what is good in this paper and none of what is not.

The research reported in this chapter was supported by the National Science Foundation under grant no. TPE-8954724 and by the Spencer Foundation under grant no. 1996–00130. The opinions expressed do not necessarily reflect the views of either foundation.

Notes

1. Reform documents on which this characterization is based include: National Council of Teachers of Mathematics (1989, 1991); National Research Council (1988).
2. See, for example, Gamson (1992), Geisler (1994), Kuhn (1992), and Schiffrin (1984).
3. The suggestion is not that in some circumstances one may have a face and in others not; rather, the thesis is that one has different faces, and one may be more or less attached to any one face in different ways and under different circumstances. Goffman intends that the form of one's face and one's attachment to it is expectedly, indeed unproblematically, variable. Whereas, in contrast, variation in one's level of (self-)esteem is generally considered undesirable. Although *face* in its everyday usage suggests a discrepancy between surface appearance and reality, Goffman's and the present use are alike in not being intended to entail such a distinction.
4. Though Brown and Levinson use the terms *speaker* and *hearer*, Goffman's use of *recipient* is more suggestive that FTAs may be felt by others than those who are "addressed to." As this is a preliminary work, the matter of person-directed intentionality is not taken up in any detail, and the terms *speaker* and *interpreter* are adopted. An interpreter may be, as Goffman notes, the speaker himself.

5. For a discussion of the congruence between Lampert's classroom and the practices advocated in the *Standards,* see Lampert (1990).
6. See Cazden (1986) for a review of the literature.
7. See, e.g., Cazden (1988), Edwards & Furlong (1978), Lemke (1990), and Philips (1972).
8. For the purposes of this work, videotapes of interactions on one day during which all the students and Lampert are mutually recognized as participants were transcribed by the author. This amounts to about forty-five minutes of action. Which interactions to transcribe, that is, what counted as a lesson, was determined on the basis of loudness of speech, range of gaze, and the turn-taking mechanism in play. The transcripts are segmented into words, utterances (a set of words bounded by a falling intonation and sometimes by another speaker's turn), and turns (a set of utterances or action by a single speaker bounded on either side by a set of utterances or action by a different speaker), and they are presented so that the differences between words (separated by blank spaces), utterances (separated by carriage returns), and turns (preceded by the name of the speaker) are readily visible. Time codes are given in minutes:seconds from the beginning of the lesson to indicate the position of each excerpt in the sequence of the whole. The transcriptions represent standard orthography rather than an attempt to approximate the spoken word. "//" and "[]" indicate the bounds of overlapping action and author's comments, respectively. A single period "." indicates the falling tone characteristic of the end of an utterance; multiple periods indicate silence, in seconds. In the descriptions of the lesson, material enclosed in quotation marks is taken verbatim from the transcript.
9. Thom Dye is the homeroom teacher and teaches science, social studies, and language arts to the same class.
10. The method of segmentation closely follows Mehan's (1979a) method of determining a Topically Related Set. In the Lampert classroom transcript, a Topically Related Set is delineated by Lampert's use of "now," or "suppose," or "what if."
11. Strictly speaking, the risk for face is merely the potential for damage. For the remainder of the discussion, to achieve brevity and clarity in presentation at a possible cost of oversimplification, it will be assumed that the possibility of threat is actual.
12. See Brown and Levinson, pp. 125–129.
13. Many researchers (e.g., Lemke, 1990) have noted that this teacher move, considered the third part of the IRF or IRE sequence, frequently does more than simply indicate whether the teacher counts the student's answer as "right" or "wrong." By commenting on a selected aspect of that answer, the teacher move can be considered not just "evaluation," but "elaboration" or "formulation." Mehan (1979a) argues that the underlying three-part norm of interaction in lessons ensures that a teacher's act following a student response is interpreted as evaluative *even when the evaluation is not (overtly) linguistically realized.* In Goffman's (1956) view as well, inaction (a null move, in Mehan's example) carries meaning just as saying "good" or "sorry, next" would.
14. See Heidegger (1962) and Garfinkel (1967) for discussions of the phenomenology and significance of breakdown.
15. As Mehan (1979b) and many others have observed, questioning in typical classrooms is frequently pseudo-questioning, involving questions whose answers are already known to the questioner. In everyday circumstances, questioning is different in that by indicating a presumption on the part of the speaker that the interpreter has

the information sought by the speaker, it defeats the presumption that the speaker already has the information he is asking for. Thus, true questioning of what a student means presupposes that the teacher does not already know what a student intends by his words; insincere or pseudo-questions presuppose that the teacher does know what is meant.

References

Brown, P., & Levinson, S. (1987). *Politeness: Some universals in language usage.* New York: Cambridge University Press.

Cazden, C. B. (1986). Classroom discourse. In M. C. Wittrock (Ed.), *Handbook of research on teaching* (3rd ed.), pp. 432–463. New York: Macmillan.

Cazden, C. B. (1988). *Classroom discourse: The language of teaching and learning.* Portsmouth, NH: Heinemann.

Cohen, D. K. (1988). *Teaching practice: Plus ça change . . .* (Issue Paper 88–3). East Lansing, MI: Michigan State University, The National Center for Research on Teacher Education.

Edwards, A. D., & Furlong, V. J. (1978). *The language of teaching.* Portsmouth, NH: Heinemann.

Gamson, W. (1992). *Talking politics.* New York: Cambridge University Press.

Garfinkel, H. (1967). *Studies in ethnomethodology.* Englewood Cliffs, NJ: Prentice-Hall.

Geisler, C. (1994). *Academic literacy and the nature of expertise: Reading, writing, and knowing in academic philosophy.* Hillsdale, NJ: Erlbaum.

Goffman, E. (1955). On face-work: An analysis of ritual elements in social interaction. *Psychiatry, 18,* 213–231.

Goffman, E. (1956). The nature of deference and demeanor. *American Anthropologist, 58,* 473–502.

Goffman, E. (1981). Replies and responses. In E. Goffman, *Forms of talk,* pp. 5–77. Philadelphia: University of Pennsylvania Press. (Reprinted from *Language in Society, 5,* 257–313, 1976.)

Heidegger, M. (1962). *Being and time.* New York: Harper & Row.

Jackson, P. W. (1968). *Life in classrooms.* New York: Holt, Rinehart & Winston.

Kuhn, D. (1992). Thinking as argument. *Harvard Educational Review, 62* (2), 155–178.

Lampert, M. (1990). When the problem is not the question and the solution is not the answer: Mathematical knowing and teaching. *American Educational Research Journal, 27,* 29–63.

Lampert, M., Rittenhouse, P., & Crumbaugh, C. (1996). Agreeing to disagree: Developing sociable mathematical discourse in school. In D. R. Olson & N. Torrance (Eds.), *Handbook of education and human development,* pp. 731–764. Oxford: Basil Blackwell.

Lemke, J. L. (1990). *Talking science: Language, learning, and values.* Norwood, NJ: Ablex.

Mehan, H. (1979a). *Learning lessons: Social organization in the classroom.* Cambridge, MA: Harvard University Press.

Mehan, H. (1979b). "What time is it, Denise?": Asking known information questions in classroom discourse. *Theory into Practice, 18* (4), 285–294.

Morine-Dershimer, G. (1983). Instructional strategy and the "creation" of classroom status. *American Educational Research Journal, 20* (4), 645–661.

National Council of Teachers of Mathematics. (1989). *Curriculum and evaluation standards for school mathematics.* Reston, VA: National Council of Teachers of Mathematics.

National Council of Teachers of Mathematics. (1991). *Professional standards for teaching mathematics.* Reston, VA: National Council of Teachers of Mathematics.

National Research Council. (1988). *Everybody counts: A report to the nation on the future of mathematics education.* Washington, DC: National Academy Press.

O'Connor, M. C., & Michaels, S. (1993). Aligning academic task and participation status through revoicing: Analysis of a classroom discourse strategy. *Anthropology and Education Quarterly, 24* (4), 318–335.

Philips, S. (1972). Participant structures and communicative competence: Warm Springs children in community and classroom. In C. Cazden, V. John, & D. Hymes (Eds.), *Functions of language in the classroom.* New York: Teachers College Press.

Pontecorvo, C., & Girardet, H. (1993). Arguing and reasoning in understanding historical topics. *Cognition and Instruction, 3-4* (11), 365–395.

Schiffrin, D. (1984). Jewish argument as sociability. *Language in Society, 13* (3), 311–335.

Afterword

10 Closing Reflections on Mathematical Talk and Mathematics Teaching

Deborah Hicks

The reform movements in the field of mathematics education are ones that, to those outside the discipline, seem to cut to the core of how mathematics is taught in classrooms. As an educator who studies how children learn in classrooms and also works with preservice and inservice teachers, I have experienced the impact of new math standards for teaching and evaluation. Teachers with whom I conduct classroom research are now beginning to take seriously the mandate that children need to communicate mathematically if they are to participate fully in mathematics learning. Magdalene Lampert articulates succinctly the pedagogical belief that students must be able to participate in authentic mathematics discourses if they are to learn to reason mathematically. In her words (this volume, p. 10), "If they [students] are to conjecture and connect, they will need to communicate." The published mathematics education standards, and their use among preservice and inservice teachers, has begun to result in changes evident to those of us working directly in classrooms. Even children in kindergarten classrooms are being asked to articulate their mathematical reasonings. Moreover, mathematical discussions are being used as ways to scaffold children's reasonings. The role of communication is being turned on its head in some reform classrooms. It is being used as a critical tool in the construction of knowledge, not simply an articulation of what is assumed to be already "inside" the learner's head.

In these closing reflections I want to build on the work of the various scholars contributing to this volume both to celebrate these enormous changes in mathematics research and practice and to look at some still neglected topics and concerns. I write these reflections from the perspective of one who stands on the periphery of the work described in this volume. Although I have written extensively on the role of discourse in the classroom construction of knowledge (see Hicks, 1995–96, 1996), I am not a mathematics educator. Rather, I work most closely in the area of literacy education, looking at a broad range of issues that impact children's reading, writing, and their talk in classrooms. However, I view this vantage point as one beneficial to the work at hand. As an outsider peering into the work described in this volume, I can be both admiring of its enormous merits and also cognizant of what work might still need to be done. In fact, I will draw on some insights that have been gained from studies of literacy, connecting these to the kinds of questions posed in this volume. For instance, some of the deepest insights that have been gained from decades of research on language (and its connections to literacy) relate to the wide differences in language use across communities. The question of "equal access" to mathematical forms of reasonings emerges as a critical one in that regard. Also, the teaching of discipline-specific discourses, the *genres* of subjects like mathematics, has been a hotly debated one among literacy educators. I will draw on my knowledge of topics like these as I reflect on the work accomplished in this volume. First, however, I want to highlight some of the issues raised by Lampert in her introduction and explored more fully in the individual chapters.

In her introductory comments, Lampert discusses three bodies of scholarship important to the work described in this volume: scholarship on mathematics as a set of disciplinary practices and beliefs mediated by specific discourses, work on the nature of knowledge and how it is constructed in social settings, and finally scholarship on the teaching practices that take place in classrooms. These three bodies of scholarship are ones that will frame my own closing thoughts on the kinds of contributions made by the collective work in this volume. This work might mistakenly be viewed primarily as exemplars of mathematics education teaching and research in which discourse is taken seriously. It certainly is noteworthy on that score alone, since exemplars of just how discourse is to be woven into authentic mathematics teaching and learning are few and far between. Although educators now generally accept the premise that communication is essential for any disciplinary teaching and learning, just *how* talk is to be used effectively is less clear. Thus, a legitimate function of a volume like this one is its compilation of exemplars both of how mathematical talk can be accomplished in actual classroom practice and how mathematical talk can be studied by researchers. A less obvious contribution of a volume like the present one, however, is its complex weaving of new epistemologies of mathematics teaching and learning, ones that relate specifically to the role of talk and other social practices. It is worth a short detour, then, to recall how mathematical discourses, mathematical understandings, and teaching practices are interwoven pedagogical strands.

Many of the contributors to this volume draw judiciously on a theory of learning that is rooted in the social nature of knowledge construction. The intellectual debt to the work of scholars like L. S. Vygotsky is evident throughout, as the contributors have explored the ways in which social discourse is part of the knowledge constructed among participants. It is a serious shift in perspective to consider overt forms of intermental activity as fundamental to the study of learning – in this case overt mathematical activity like *talk.* One way to characterize the theoretical strands developed in this volume (with "theory" certainly inclusive of actual classroom practices and their underlying philosophies) is to suggest that all of the chapters, in some way, touch upon *mathematical literacies.* In using the term *literacies,* I am drawing on its usage by scholars like Gee (1990) and Michaels, O'Connor, and Richards (1993): literacies as forms of reasoning embedded in disciplinary discourses, belief systems, and social practices. Authentic mathematical literacies would then, as is discussed in the various chapters, include much more than procedural knowledge – the "how to" that is the hallmark of traditional mathematics teaching and learning. They would resemble more the kinds of discourses and interactions demonstrated among the fifth-grade students in Lampert's classroom, as these unfolded after extensive teacher scaffolding. In various ways and among differing age groups (even including professional engineers in a work situation), contributors to the volume explore what mathematical literacies are (chapter 2), how they emerge through interactions with teachers and other students (chapters 5, 7, 8, and 9), and how costly it is to students when they are not constructed in small group settings (chapter 4).

Thus, the shift in perspective from looking at mathematics learning as internal reasoning processes to looking at what is *interactionally accomplished* through talk is a crucial one to note. Certainly, the contributors to this volume are concerned with students' constructing deep, generative understandings of mathematical ideas. And yet, the methods for studying mathematical teaching and learning proceed, as in Vygotsky's famous critique of Piagetian egocentrism (see Vygotsky, 1986), from the opposite direction: from what is socially experienced and talked about to what later becomes intramentally generative for the individual learner. Research on mathematics teaching and learning can now be studied through the analysis of transcripts of talk and interaction. Literacies can be positioned by teachers and researchers as central to mathematical knowing.

The question for mathematics educators then becomes one of just how such authentic mathematical literacies are inculcated in the classroom, surely a social setting like no other, where participation structures and forms of talk are often highly regulated and often not all that much like real disciplinary practices. Numerous chapters address just that question, looking at how the construction of mathematical *literacies* (again, in the broader sense of mathematical discourses and forms of knowing) can be accomplished among students and teachers. What the various chapters reveal is that the social accomplishment of mathematical discourses and ways of knowing in classrooms – what various contributors refer to as *authentic* mathematical activity – is both possible and challenging. The construction of mathematical conversations and their subsequent analysis, described in part II of this volume, shows the complexity of the whole process. Deliberate participation structures, modes of talk, and belief systems were supported by the teacher-researcher (Lampert) across the year among fifth graders. Multiple agendas were juggled as children negotiated social roles and relationships while they learned to participate in mathematical discourses. Beliefs about what it means to think and act mathematically were changed as students were apprenticed into new kinds of discourse – ones that showed a willingness to explain and justify one's ideas but also a responsiveness to the ideas of others in the classroom. The differing interpretive perspectives brought to this long-term project (see chapters 6–9) reveal the intricate ways in which such mathematical literacies are constructed among participants in classrooms. The work presented in chapter 5 also shows through microanalysis of one episode how a student and tutor co-construct mathematical forms of reasoning through discourse; in short, how mathematical literacies are socially accomplished.

The notion of mathematical literacies as forms of talk and forms of reasoning is therefore an important strand that undergirds the volume, in that it links emphases on theory, research, and practice. And yet, there is even more to be said about mathematical literacies, *if* we think of these as "identity toolkits" (Gee, 1990), part of who we are as persons in relationship in classrooms and communities. It is to that subject I now turn in this latter part of my closing reflections.

Implications of Research on Language and Literacy

The subfield of research on language and literacy education has been profoundly influenced by studies of the sociocultural underpinnings of language learning. The insights that have been gained through decades of sociocultural and sociolinguistic research merit some attention in the present context, since this type of research has yet to have a similar impact on mathematics educators. The teaching and evaluation standards published by the National Council of Teachers of Mathematics have as one goal the accordance of equal access to mathematical knowledge. Bringing mathematics out of the dark, so to speak, and more into overt processes of communication and small group work, should afford more children greater access to mathematical literacies. Like many scholars looking admiringly on the changes occurring in the field of mathematics education, I agree with this new emphasis on communication. At the same time, I am cognizant of the findings from numerous studies of relations between home, community, and school discourses. The research that has been developed among literacy educators has revealed important relationships between the "ways with words" (Heath, 1982, 1983) of communities and classrooms, ones that profoundly impact how children engage with formal, academic literacies. Although much of this work has been applied to the thorny dilemma of access to *literacy* in the more traditional sense of the word (i.e., reading and writing), this work could potentially be highly informative for studies of mathematical literacies.

Rather than make an attempt to review this enormous body of scholarship (see Hicks, 1996, for just such a review), which would be much too cumbersome for my purposes here, I will cite just one exemplar. In what has become something of a classic in the field of language and literacy education, anthropologist Shirley Brice Heath spent the better part of a decade studying patterns of language socialization among three communities in the Piedmont area of the eastern United States. Through close and long-term ethnographic documentation, including the extensive collection of language data (e.g., stories, conversations), Heath (1982, 1983) uncovered distinctly different patterns of language learning and use among communities in fairly close proximity to one another. Heath's emphasis in this long-term ethnography was on the *discourses* of the three communities she studied – how participants used language socially in the settings that made up their community lives. In ways critical for understanding children's access to formal school literacies, members of the three communities were participants in different kinds of discourses. For instance, young members of a working-class black community participated in vibrant storytelling and conversational episodes that often melded fact and fiction. Even very young children became accomplished at nonliteral meanings and verbal repartee. However, these same children did not participate in school-like conversations in which an adult participant "froze" an activity through narrative description or explanation.

Moreover, they did not typically experience bedtime book readings, where they would become apprentice users of literary discourses.

What Heath found through her collaborative work with teacher researchers (who were themselves typically members of a nearby middle-class community) was that *deliberate* efforts to take into account children's sociocultural and linguistic backgrounds were needed before children experienced more equal access to school knowledge. Teachers in Heath's study became co-researchers who studied the language and social backgrounds of their students and then developed curricula that mediated between differing discourses and worldviews. Teachers themselves became ethnographers of communication, developing first a deeper understanding of the patterns of language socialization unique to the groups of students in their classrooms. The teaching practices described by Heath were oriented toward helping children achieve literacy in the broader sense of forms of reasoning embedded in certain kinds of communication. For instance, middle-grade science students researched the scientific discourses of their communities and then learned about the more formal academic discourses used by scientists. Heath's efforts to help teachers mediate between the "ways with words" of communities and classrooms parallel other efforts to connect students' home and school lives (see, for instance, the work of Ballenger [1994] on teaching science to Haitian students; and the work of Luis Moll and his colleagues [e.g., Moll & Dworin, 1996] on teaching literacy to children living in Latino *barrios*).

Studies of relations between students' community and classroom discourses have yet to have a similarly important impact among mathematics educators. The literacies that are now so skillfully modeled in some mathematics classrooms are assumed to be outside the experience of most children. As contributing author Rittenhouse states (see chapter 7), they are something to be learned, not acquired. And yet, the extensive body of sociocultural and sociolinguistic research to which I have alluded suggests that children may bring to the classroom vastly different degrees of familiarity with certain *kinds* of discourses that in important ways resemble mathematical literacies. If children have not experienced discourses that, in quite deliberate ways, "break up" or "freeze" the action occurring, then the explicit articulation of one's reasoning processes may seem quite daunting as a form of social practice.

This very difficult issue runs even deeper than I have thus far intimated, since it cuts to the very core of "who we are" as persons in relationship, embracing values and beliefs that typically reflect those of our homes and communities. In the theorizing and research that has occurred subsequent to some of the earlier sociolinguistic studies of communities (such as Heath's seminal work), scholars have drawn on a wide range of literatures to address just how deeply implicated discourses are in the fabric of our lives. Theorists like Michel Foucault have written about how discourses structure the work of institutions, and how disciplinary knowledge and practice is embedded in what he calls discursive formations (e.g., Foucault, 1972). Educational theorists have drawn on these critical studies of discourse to write about the ways in which discourses embody the activities, ways of using language, and values and beliefs of certain kinds of people. My participation in a discourse is both participation in a set of language practices (that are also of course connected to larger social practices), and also a reflection of the kind of person that I am. An example that I might use based upon the writings of James Paul Gee is the following. If I walk into a motorcycle bar, I (a female middle-class academic) might speak the same language, but I don't speak the discourse. I would be immediately recognized as a nonmember of that social setting. Gee (1990) and other scholars sometimes use uppercase *D* (*Discourse*) to indicate the term's larger meaning as a set of language practices inclusive of values, beliefs, and ways of acting. As I suggested earlier, these are "identity toolkits," not simply language genres.

This more inclusive interpretation of what discourses are bears direct relevance to the study of mathematics teaching and learning, for mathematical forms of talk are also Discourses. They reflect certain belief systems and values, and they ultimately convey a sense of "who students are" as persons in relationship to home and school. Children may experience difficulty with mathematical literacies, not so much because they cannot learn linguistic structures, but because of the conflicting alignments that discourses entail. There is a danger in thinking of discourses as textual structures that are "out there," divorced from the persons who are appropriating them. Such a view might imply that if students can simply learn to communicate mathematically, they will be able to participate fully in the forms of reasoning that characterize the discipline. The philosopher and literary theorist Mikhail Bakhtin was adamant in linking utterances of any sort (spoken or written) to both stable discourse genres and centers of value (e.g., Bakhtin, 1981, 1993). Weingrad (chapter 9) makes a similar point when she suggests that the mathematics teaching exemplified in part II of this volume is implicated in moral decisions and acts. Students place themselves "on the line" in various ways as they socially position themselves with respect to formal literacies and to their peers and teachers. The conflicts they may experience in doing so may be simultaneously conceptual and moral – in the sense that they are appropriating forms of talk and practice that seem alien to who they are. Students may resist, conveying through words or actions their alignment with other Discourses. Teachers who work with students from non-middle-class backgrounds may find that teaching formal mathematical literacies demands a consideration of language and cultural differences as well as a sensitivity to the possibly conflicting values that these may entail.

It might be valuable at this point in my discussion to make note of some reform efforts in Australia and New Zealand that place at center stage the teaching of formal literacies. The *genre movement* (as it has become known among literacy educators) has been a direct effort to reform teaching and learning around the idea that specific genres structure academic disciplines and that domain-specific teaching must include explicit genre instruction (see Cope & Kalantzis, 1993; Reid, 1987). Such instruction might, in a mathematics educational setting, include things like teaching students how to construct a mathematical narrative (Longo, 1994) or explanation. To some extent, such deliberate "genre instruction" is part of what happens at least implicitly in reform mathematics classrooms. Teachers deliberately call students' attention to the forms of communication appropriate for group discussions (see the discussion in part II of this volume). The genre instruction movement in Australia and New Zealand goes even further than this in weaving explicit grammar instruction (that is, the *text grammars* of academic discourses) into the teaching of subjects like social studies and science. What we in this country would refer to as language arts instruction is woven into discipline-specific teaching. In the context of learning about the greenhouse effect, for instance, students receive grammar instruction (e.g., modeling, structured practice sessions) on how to write a scientific explanation [see Callaghan, Knapp, & Noble, 1993]). This pedagogical reform movement is referred to by its advocates as a pedagogy of "inclusion and access" (Cope & Kalantzis, 1993). It is meant to afford students from a wide range of cultural and economic backgrounds access to formal academic literacies.

In spite of its many merits, the genre instruction movement does not in my view adequately address how discourses (or Discourses) function as embodied expressions of values and beliefs. Theories of mathematics teaching that prioritize teaching mathematical forms of talk, for instance, might run the risk of positioning discourses as what the philosopher Amelie Rorty refers to as "runes and ruins" – artifacts that represent the social practices typical of accomplished mathematicians (see Rorty, 1995). As I read the many examples of reform mathematics teaching in this volume, they entail a sensitivity both to what counts as "authentic" mathematical discourse and to the struggles that students experience in giving voice to those mathematical discourses. Thus, as mathematics educators grapple with what it means to learn mathematical literacies, they might refer to theories of language that are inclusive of values and beliefs. Once we have collectively identified the authentic discourses of the mathematics community and then talked about how these can be most effectively taught in the classroom, there is still the day-to-day world of teaching in which multiple agendas merge, or collide.

A useful way of thinking about mathematical literacies might be to reflect on them as instructional processes rather than end products to be "attained" by students. The research described in chapter 3, for instance, documents the recursive nature of the instruction that occurred in a first-grade classroom. Good mathematics instruction at all grade levels is now (at least among reform mathematics educators) assumed to involve recursive problem solving that may or may not entail getting the "right answer" (Lampert, 1990). Similarly, mathematical literacies might be viewed as activities that we would hope to be generative ones for the individual learner. Participant observation of the kind described in many chapters in this volume allows researchers and teachers to study the processes by which students and teachers co-construct mathematical talk. It would seem to be these social and discursive processes that are most important in mathematics teaching and learning, not the disciplinary genres themselves. Drawing again on Amelie Rorty's writings on reading as an act of cultural interpretation, we could say that literacies as cultural texts, practices, and beliefs are "private and collective . . . '*workings through*' " that ultimately involve the learner as an active agent and author (Rorty, 1995, p. 221, emphasis added). In all the recent excitement over authentic communication as a conduit to mathematics learning, it is all too easy to lose sight of how students need to do more than "talk the talk." They also need to use mathematical literacies in generative, interpretive ways and to place some personal value on mathematical discourses. But there is the rub: Literacies as personal and collective "workings through" are difficult to teach in classrooms. Moreover, a lot depends on the expertise of teachers, who must themselves understand what it means to "work through" mathematical literacies in a more generative sense.

Thus, I have returned full circle to some of the opening concerns that coeditor Lampert raises in her introduction, with her emphasis on the *practice* of mathematics teaching and learning. As an educator who works with preservice and inservice teachers *and* who studies children's learning in classrooms, I am painfully aware of the challenge of creating teaching environments where all, or even most, students experience mathematical literacies in the ways we would like. In thinking about how mathematical literacies might be inculcated on a wider scale, I am once again drawn to the discipline with which I am more intimately familiar, the field of literacy education. In response to some of the more progressive movements in literacy education, teachers and literacy researchers have formed a wide range of support systems for literacy educators. Even before progressive movements such as the "whole language" movement became part of our popular educational vocabulary, institutions like the Breadloaf Writer's Conference (a summer "camp" of sorts for serious writers) sponsored special workshops for language arts teachers. Teacher researchers like Nancie Atwell (see Atwell, 1987) found their footing, so to speak, in the challenging and supportive context of the Breadloaf conference, using time spent among serious writers to rethink their teaching practices. The National Writing Project is another example of a place where language arts teachers converge to think critically about their practice. In the summer, teachers converge for several weeks to participate in dialogues about writing and literacy, under the apprenticeship of both classroom teachers and university educators. In these and other more local contexts for critical dialogue, classroom teachers experience the opportunity to step back from their practice and experience literacy as a personal and collective "working through."

In my work with classroom teachers around the topic of literacy education, I have encountered smaller scale versions of these more highly organized support networks. Individual teachers might themselves write memoirs, fiction, or poetry, and they might be members of writers' or readers' groups. Teachers might be avid readers of literature, having themselves the frequent experience of becoming immersed in a piece of fiction or poetry. Individual teachers might be members of families or social networks that engage in "dining room table talk" (Atwell, 1987) that involves literature or nonfiction writing. In these kinds of local and at times informal settings, teachers have the chance to become authors or critics of literary texts. Literacies in the ways they are manifested in language arts teaching are widely accessible and widely valued in a sense that may not be true for mathematical literacies. If mathematical forms of reasoning and talk are not something deeply valued as a cultural practice among teachers, then it would seem all that much more difficult for these literacies to be scaffolded in classrooms. Clearly, some teachers do express a deep interest in and concern for mathematics teaching and learning. The first-grade teacher in chapter 3 and the teacher researchers in chapters 4 and 6 of this volume are examples of this kind of valuing. However, a web of professional support that extends into our wider cultural lives may not be as strong a part of the mathematics education community. Perhaps, then, mathematical literacies are something that could become a stronger part of our teacher education programs. As a first step, could the kind of *epistemology* of mathematics teaching exemplified in this volume be woven into the preservice education of teachers? Could there be supportive groups that parallel the work of the National Writing Project?

I leave these questions to be addressed by those working more directly in the mathematics education community, for surely questions like these are already being pondered. However, there is still room for new kinds of dialogues among mathematics educators, perhaps ones that interface with scholars from other communities (such as literacy educators). Some of the most penetrating questions regarding language and its relation to classroom learning have yet to be critically addressed in the mathematics education field. That field has relied in great part thus far on the very hopeful view that students who communicate mathematically will have greater access to mathematical forms of knowledge. I concur wholeheartedly with that hopeful vision. However, from my vantage point as a scholar of discourse and its relation to learning, I also hold on to a healthy bit of skepticism. More richly informative, "thicker" theories of discourse need to be woven into the scholarly work of the mathematics education community, and these need to be combined with classroom research studies that utilize them in creative and productive ways. Perhaps this can best occur in dialogue with those working more directly with studies of discourse. The contribution of O'Connor (chapter 2) to this volume attests to the many possibilities of such a dialogue. Topics like discourse and its relation to identity construction, narratives (i.e., stories) and their connections to self and community, and disciplinary genres as tools of social empowerment are ones that are hotly debated in conferences and journals concerned with language and literacy. It would be an exciting development to see more of these kinds of topics threaded into the discourses of the mathematics education community.

References

Atwell, N. (1987). *In the middle: Writing, reading, and learning with adolescents.* Portsmouth, NH: Heinemann.

Bakhtin, M. M. (1981). *The dialogic imagination: Four essays by M. M. Bakhtin.* M. Holquist (Ed.), C. Emerson & M. Holquist (Trans.). Austin: University of Texas Press.

Bakhtin, M. M. (1993). *Toward a philosophy of the act.* V. Liapunov & M. Holquist (Eds.), V. Liapunov (Trans.). Austin: University of Texas Press.

Ballenger, C. (1994, February). *Heterogeneity in science talk: Argumentation in a Haitian bilingual classroom.* Paper presented at the Ethnography in Education Research Forum, University of Pennsylvania, Philadelphia.

Callaghan, M., Knapp, P., & Noble, G. (1993). Genre in practice. In B. Cope & M. Kalantzis (Eds.), *The powers of literacy: A genre approach to teaching writing,* pp. 179–202. London: The Falmer Press.

Cope, B., & Kalantzis, M. (Eds.). (1993). *The powers of literacy: A genre approach to teaching writing.* London: The Falmer Press.

Foucault, M. (1972). *The archeology of knowledge and the discourse on language.* A. M. Sheridan Smith (Trans.). New York: Pantheon Books.

Gee, J. P. (1990). *Social linguistics and literacies: Ideology in discourses.* London: The Falmer Press.

Heath, S. B. (1982). What no bedtime story means: Narrative skills at home and at school. *Language in Society, 11,* 49–76.

Heath, S. B. (1983). *Ways with words: Language, life, and work in communities and classrooms.* New York: Cambridge University Press.

Hicks, D. (1995–96). Discourse, learning, and teaching. In M. Apple (Ed.), *Review of Research in Education, 21,* 49–95.

Hicks, D. (Ed.). (1996). *Discourse, learning, and schooling.* New York: Cambridge University Press.

Lampert, M. (1990). When the problem is not the question and the solution is not the answer: Mathematical knowing and teaching. *American Educational Research Journal, 27,* 29–63.

Longo, P. (1994, February). The dialogical process of becoming a problem solver: A sociolinguistic analysis of implementing the NCTM standards. Paper presented at the Ethnography in Education Research Forum, University of Pennsylvania, Philadelphia.

Michaels, S., O'Connor, M. C., & Richards, J. (1993). Literacy as reasoning within multiple discourses: Implications for restructuring learning. In *Restructuring learning.* 1990 Summer Institute Papers and Recommendations by the Council of Chief State Officers. Washington, DC: CCSSO, 107–122.

Moll, L., & Dworin, J. (1996). Biliteracy development in classrooms: Social dynamics and cultural possibilities. In D. Hicks (Ed.), *Discourse, learning, and schooling,* pp. 221–246. New York: Cambridge University Press.

Reid, I. (Ed.). (1987). *The place of genre in learning: Current debates.* Deakin University: Centre for Studies in Literary Education.

Rorty, A. (1995). Runes and ruins: Teaching reading cultures. *Journal of Philosophy of Education, 29* (2), 217–222.

Vygotsky, L. S. (1986). *Thought and language.* A. Kozulin (Trans.). Cambridge, MA: MIT Press.

Author Index

Subject Index